The Infinite

The Problems of Philosophy
Their Past and Present

General Editor: Ted Honderich
Grote Professor of the Philosophy of Mind and Logic
University College, London

Books already published:

The Infinite

A.W. Moore

Routledge
London and New York

First published 1990
by Routledge
11 New Fetter Lane, London EC4P 4EE

Simultaneously published in the USA and Canada
by Routledge
a division of Routledge, Chapman and Hall, Inc.
29 West 35th Street, New York, NY 10001

© A. W. Moore

Typeset in 10/12 Times by J&L Composition Ltd, Filey, North Yorkshire
Printed in Great Britain by T.J. Press (Padstow) Ltd, Padstow, Cornwall

British Library Cataloguing in Publication Data

Moore, A. W.
The infinite.—(The problems of philosophy: their
past and present)
1. Infinity
I. Title
111'.6

ISBN 0–415–03307–1

Library of Congress Cataloguing in Publication Data

Moore, A. W.
The infinite / A.W. Moore.
p.   cm. — (The Problems of philosophy)
Bibliography: p.
ISBN 0–415–03307–1
1. Infinite. I. Title. II. Series: Problems of philosophy
(Routledge (Firm))
BD411.M59 1989
111'.6—dc20   89–10458

For my parents

Contents

Contents

The Infinite

Preface

Light travels at a speed of approximately 186,000 miles per second: 186,000 miles is over seven times the circumference of the earth. On a clear, moonless night a faint patch of light can be seen in the constellation Andromeda. This is the Andromeda nebula. It is a galaxy of about a hundred thousand million stars, each of them a sun like our own. Its light takes some two million years to reach us. It is the farthest object visible to the naked eye. Yet by comparison with other galaxies it is a close neighbour.

Who can fail to be moved by the sheer scale of it all? Who can deny the humbling and awesome effect of contemplating those vast, silent reaches?

It is the same when we think about the past. Imagine that the five-thousand-million-year history of the earth were condensed into a decade. Then dinosaurs died out between two and three months ago; about a fortnight ago proconsul apes appeared; nine hours ago humans were beginning to make tools; approximately two minutes ago Jesus Christ was born; and three seconds ago the atom bomb was exploded.

Yet we are still unmistakably in the realms of the *finite*. The two million light-years separating us from the Andromeda nebula, and the one hundred and fifty million years separating us from the dinosaurs: these are finite 'bits' of space and time. The infinite seems, not bigger, or at any rate not just bigger, but of a different *kind*.

This is a book about the infinite. It is also a book about the finite – about our own finitude. A sense of our own finitude is what underlies our sense of the infinite. We know that we are finite. This is not just a matter of our being tiny, and ephemeral. There is something more fundamental than that. There is the fact that we find ourselves cast into a world that is not of our own making, the fact that we find ourselves confronted with what is *other* than us. And if scientific investigation should reveal that all the suns, and planets, and meteors were contained in a finite region of space, and that they were all debris from some cosmic explosion that took place finitely many years ago – nay, that space and time themselves were finite – still we should have a contrasting sense of the infinitude that surrounds us:

the all-encompassing, unified whole. It is at that sense, always poignantly bound up with self-consciousness about our own finitude, that the majesty of the universe most persistently tugs. What are we to make of our lives – what, indeed, are we to make of our deaths – when set against the stars?

It is questions such as these that provide the backcloth for this book. My aim, in general terms, is to make sense of the infinite. I draw on what western philosophers have thought about the infinite ever since they first began to pay it attention some two and a half thousand years ago. By first outlining the history of their thought, I attempt to construct a coherent picture from the insights that they have passed on.

Inevitably, much of the book is mathematical. Nevertheless, the degree of mathematical knowledge presupposed is minimal. A glossary is included at the end of the book with basic definitions of the technical terms that occur most frequently, together with references to the sections where they are first introduced.

I should like to thank Ted Honderich for inviting me to write this book. I am also very grateful to the provost and fellows of King's College, Cambridge, who elected me into a junior research fellowship and thus provided me with the ideal opportunity to carry out the bulk of the work on it. Thanks are also due to the editors and publishers of *Cogito*, *Mind*, and *The Proceedings of the Aristotelian Society* for permission to re-use material from those journals; I have located such material at the relevant points in the text.

I cannot mention everybody who has influenced my thinking about these issues, or who has helped me, either directly or indirectly, with the writing of the book. But it is a pleasure to record specific debts to Joseph Melia and Colin Sparrow, and a very special debt to my parents.

There are two people I should like to single out for particular mention. Philip Turetzky has had a greater influence than anybody else on my philosophical thinking. He taught me to love and respect the history of philosophy; he showed me what it is to think about a philosophical problem. Naomi Eilan has been more closely involved than anybody else with my work on this project. Her continual encouragement, enthusiasm, and penetrating advice have helped me in ways that she could never know. To both of them I extend my warmest thanks. The highest compliment that I can pay either of them is to bracket them together in this way.

A.W. Moore

Two things fill the mind with ever newer and increasing admiration and awe, the more often and the more steadily they are reflected upon: *the starry heaven above me and the moral law within me*. I should not search for them and merely conjecture them as though they were veiled in darkness, or were in the transcendent region beyond my horizon; I see them before me and associate them directly with the consciousness of my own existence. The first begins at the place which I occupy in the external world of sense, and broadens the connection in which I stand into the unsurveyable magnitude of worlds beyond worlds and systems of systems, and moreover into the limitless time of their periodic motion, its beginning and continuation. The second begins at my invisible self, my personality, and depicts me in a world which has true infinity, but which is traceable only by the understanding, and with which I recognize myself as being not as before in a merely contingent connection, but in a universal and necessary one (as I thereby also am with all those visible worlds). The first view of a countless multitude of worlds annihilates as it were my importance as that of an *animal creature*, which must give back to the planet (a mere speck in the universe) the matter of which it was formed, after it has been provided for a short time (we know not how) with vital power. The second, on the other hand, infinitely raises my worth as that of an *intelligence* by my personality, in which the moral law reveals to me a life independent of animality and even of the whole world of sense, at least as far as can be inferred from the destination assigned to my existence by this law, a destination which is not restricted to the conditions and limits of this life, but reaches into the infinite.

(Immanuel Kant)

Introduction: Paradoxes of the Infinite

The infinite has always stirred the emotions of mankind more deeply than any other question; the infinite has stimulated and fertilized reason as few other ideas have; but also the infinite, more than any other notion, is in need of clarification. (David Hilbert)

The aim of this book is to arrive at an understanding of the infinite – *via* an understanding of how it has been understood by other thinkers in the west over the past two and a half millennia.

It would be inappropriate to try to begin with a crisp, substantive, uncontroversial definition of the infinite. There are two special reasons for this. First, one of the central issues concerning the infinite is whether it *can* be defined. Many have felt that it cannot; for if we try to define the infinite as that which is thus and so, we fall foul of the fact that being thus and so is already a way of being limited or conditioned. (It is as if the infinite cannot, by definition, be defined. This is one of the paradoxes that we shall be looking at later in this introduction.) Despite this, there have been many attempts throughout the history of thought about the infinite to define it, or at least to explain why it cannot be defined by those persuaded that it cannot. And these supply the second reason why it would be inappropriate, in a book where historical impartiality at the outset is crucial, to try to begin with a preferred definition: these attempts have revealed a striking lack of consensus. It is not just that different thinkers have focused on different aspects of the infinite. Again and again we find new accounts of the infinite being presented in the firm conviction that what had been handed down as orthodoxy was just wrong.

Two clusters of concepts nevertheless dominate, and much of the dialectic in the history of the topic has taken the form of oscillation between them. Within the first cluster we find: boundlessness; endlessness; unlimitedness; immeasurability; eternity; that which is such that, given any determinate part of it, there is always more to come; that which is greater than any assignable quantity. Within the second cluster we find: completeness; wholeness; unity; universality; absoluteness; perfection; self-sufficiency;

1

autonomy. The concepts in the first cluster are more negative and convey a sense of potentiality. They are the concepts that might be expected to inform a more mathematical or logical discussion of the infinite. The concepts in the second cluster are more positive and convey a sense of actuality. They are the concepts that might be expected to inform a more metaphysical or theological discussion of the infinite. Let us label the concepts '*mathematical*' and '*metaphysical*' respectively.

It would be hyperbolic to say that there is no connection between the two clusters of concepts. An obvious link is the concept of being unconditioned. This could naturally be classified in either way, carrying overtones both of unlimitedness and of autonomy. Nevertheless the concepts are not obviously of a piece (which is why those philosophers who have seen the infinite in terms of one cluster have been able to accuse those who have seen it in terms of the other of being in error). There is even a hint of conflict. The concepts in the first cluster carry a sense of uncompletability, those in the second of actual completion. There may not be any deep incompatibility here. (Think about time, as a whole: it seems to be complete, but not, at any point in its midst, completable.) But still, if we are to understand the infinite, particularly if we are to understand it through its history, then one thing we must try to do is address the puzzle of why there should be this curious polarization and what exactly the concepts have to do with one another.

The puzzle is exacerbated by the fact that what we have labelled the mathematical concepts, though they do inform the most recent formal mathematical accounts of the infinite, certainly do not do so by acting as its equivalents in the way that we might have expected. Once concepts like boundlessness, or endlessness, or being greater than any assignable quantity, have themselves been made precise in various (now) standard ways, they prove to be different, one from another and each from the concept of infinity (in its own appropriately technical sense).[1] To take a simple example, the surface of the earth is not bounded, but nor is it infinite. Again, there are infinite sequences which have a bound, and there are infinite sequences which have an end (and there are some which have one but not the other); and there are infinite sets whose sizes are not only assignable quantities but smaller than other assignable quantities. Much of this is elucidated in the course of the book. It should already be clear, however, that if we are not to prejudice any issues and abrogate the very concerns and problems that are supposed to be animating this enquiry, then we must be content to start with raw, unarticulated intuitions.

The problem is that these themselves are riddled with paradoxes. I shall use this introduction to present a sample of these paradoxes. (Many more will crop up in the ensuing historical drama.) If the concept of the infinite is not ultimately to be dismissed as incoherent, then they represent the most serious threat that it faces, the abyss of absurdity from which it must be rescued. It is true that throughout the history of the topic there have been those who have looked upon the concept with suspicion, or incomprehension, or worse. But there have also proved to be continuing and irresistible pressures against eschewing it completely, felt most keenly, as often as not, by the same people. It is not a serious option to react to the paradoxes that I am about to outline by simply jettisoning the concept of the infinite as one that we are well rid of.

These paradoxes fall into four groups: paradoxes of the infinitely small; paradoxes of the infinitely big; paradoxes of the one and the many; and paradoxes of thought about the infinite. The first two groups reflect an important distinction within the mathematically infinite between what Aristotle called the infinite by division and the infinite by addition:[2] a straight line, for example, is infinite by division if between any two points on it there is a third (so there is no limit to how small a segment of the line you can take); it is infinite by addition if beyond any two points on it there is a third (so there is no limit to how large a segment of the line you can take).

1 Paradoxes of the infinitely small

(i) *The paradox of Achilles and the tortoise*

Suppose that Achilles, who runs twice as fast as his friend the tortoise, lets her start a certain distance ahead of him in a race. Then before he can overtake her, he must reach the point at which she starts, by which time she will have advanced half the distance initially separating them. Achilles must now make up this distance, but by the time he does so the tortoise will have advanced again. And so on *ad infinitum*. It seems that Achilles can never overtake the tortoise. On the other hand, given the speeds and distances involved, we can calculate precisely how long it will take him to do so from the start of the race.

Comment: This is perhaps the most celebrated and also one of the oldest of all paradoxes concerning the infinite. It is due to Zeno – if not in exactly this form. (None of Zeno's original writings on the so-called paradoxes of motion has survived. And although the tortoise appears in nearly all accounts of this paradox, going back at least as far as Simplicius, she does not appear in the earliest surviving account, in Aristotle.[3]) This paradox will be placed in its historical context later in the book (see below, Chapter 1, §3).

(ii) *The paradox of the staccato run*

Suppose that Achilles runs for half a minute, then pauses for half a minute, then runs for a quarter of a minute, then pauses for a quarter of a minute, and so on *ad infinitum*. At the end of two minutes he will have

3

stopped and started in this way infinitely many times. Yet there is something repugnant about admitting this possibility, even as a conceptual – let alone a physical – possibility. For example, suppose that each time he pauses he performs a task of some kind, there being no limit to how quickly he can do this. Then at the end of two minutes he will have performed infinitely many of these tasks. He might, say, have written down the complete decimal expansion of π (3.141592 . . .), for which he needs only a finite sheet of paper and the ability to write down digits that get smaller without limit, as Figure 0.1 testifies. We are loath to admit this as a conceptual possibility, though we seem bound to do so.

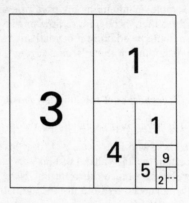

Figure 0.1

Comment: This paradox also creates unease about what would otherwise be a very natural reaction to the first paradox: namely, to insist that there is nothing incoherent in the idea of Achilles' performing infinitely many tasks in a finite time (in particular, covering the infinitely many sub-distances between his starting point and the point at which he overtakes the tortoise).

(iii) The paradox of the gods

Suppose that Achilles wants to run straight from *A* to *B* but there are infinitely many gods who, unbeknown to one another, each have a reason to prevent him from doing so. The first god forms the following intention: if and when Achilles gets half way, to paralyze him. The second god forms the following intention: if and when Achilles gets a quarter of the way, to paralyze him. And so on *ad infinitum*. All the gods are able to carry out their intentions. Achilles cannot make any progress without violating the intention of at least one of them – indeed the intentions of infinitely many of them. Yet, if he is unable to move, it is not clear why; until he makes *some* progress, none of the gods will have actually paralyzed him.

Comment: This paradox is essentially due to Benardete.[4]

(iv) The paradox of the divided stick

Suppose that an infinitely divisible stick is cut in half at some point in time, and that each half is in turn cut in half, half a minute later, and that each quarter is in turn cut in half, a quarter of a minute later than that, and so on *ad infinitum*. What will remain at the end of the minute? Infinitely many infinitesimally thin pieces? Do we so much as understand this?

Comment: Does an infinitesimally thin piece have *any* width? If so, how come infinitely many of them do not make up an infinitely long stick? If not, how can (even) infinitely many of them make up a stick with any length at all? The paradoxes that arise from envisaging the infinite division of a body were noted and discussed by Aristotle (see below, Chapter 2, §4).

2 Paradoxes of the infinitely big

We now turn to the second group of paradoxes. Consider Figure 0.2. Are there as many apples here as bananas? Or, if you like, does the set of apples have as many members – is it the same size – as the set of bananas? We can see that the answer is yes, because we can see that there are seven of each. But to see this we must count; and counting is itself an operation that presupposes such comparisons of size. To say that there are seven apples is to say that there are as many apples as there are positive whole numbers up to and including seven. (So to count the apples and the bananas is simply to bring a third set into the reckoning.)

We could, however, have answered the question from scratch, without recourse to counting – by pairing the apples and bananas off with one another, in such a way that each apple corresponds to a unique banana and each banana to a unique apple, as shown in Figure 0.3. For it to be possible to pair off the members of two sets with one another in this way seems to be what it *is* for the two sets to have as many members as each other. Applying this principle to the infinite, however, yields further paradoxes.

Before I proceed to these paradoxes I need to explain what I mean by a *natural number* and a *rational number*. (There are frequent references to these two kinds of numbers throughout the book.)

(a) The *natural numbers* are the non-negative whole numbers 0, 1, 2, ...

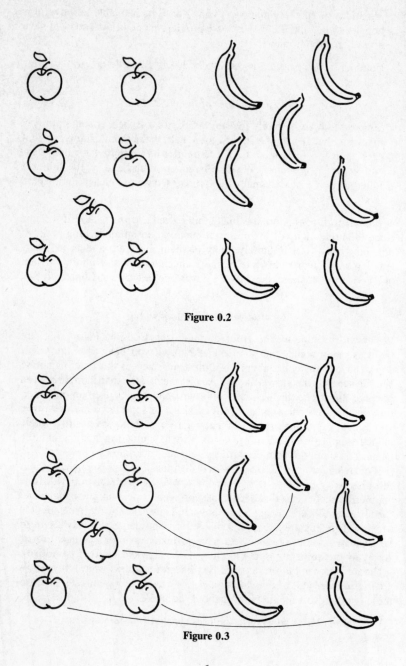

Figure 0.2

Figure 0.3

(b) The *rational numbers* (or *rationals*) are the quotients, or ratios, of whole numbers, negative and non-negative. Thus the rationals are all the numbers of the form p/q, where p and q are whole numbers and q is not 0. Examples are:

½ (= 1/2; it is also, for that matter, 2/4, 3/6, −2/−4, ...);

1½ (= 3/2);

2 (= 2/1; this, of course, is a natural number as well);

and

−1½ (= −3/2).

(i) The paradox of the even numbers

Figure 0.4 shows that we can pair off all the natural numbers with those that are even. If we apply the principle enunciated above, this shows that there are as many even numbers as natural numbers altogether. On the other hand it seems obvious that there are fewer (though we may be wary of saying that there are half as many).

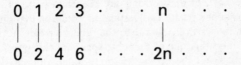

Figure 0.4

Comment: Quite apart from this paradox our intuitions here are in a state of turmoil. For even if this pairing off had not been brought to our attention, there would have been an urge to say that there are as many even numbers as natural numbers altogether; after all, there are infinitely many of each. (There is something highly counter-intuitive about the idea that one infinity can be greater than another.) It seems that however we describe the situation we shall be left feeling dissatisfied.

(ii) The paradox of the pairs

Consider Figure 0.5, in which every possible pair of whole numbers occurs once. Starting at the centre with the pair ⟨0, 0⟩ we can trace out a path as shown in Figure 0.6. Every pair eventually occurs on this path, and this is enough to show – again, counter-intuitively – that we can pair them all off with the natural numbers; for we can count as we go along.

Comment: Part of the force of this paradox, which is similar to a result established by Cantor, is that there are at least as many pairs as rationals.

$$\vdots$$

⟨−2, −2⟩	⟨−2, −1⟩	⟨−2, 0⟩	⟨−2, 1⟩	⟨−2, 2⟩
⟨−1, −2⟩	⟨−1, −1⟩	⟨−1, 0⟩	⟨−1, 1⟩	⟨−1, 2⟩
⟨0, −2⟩	⟨0, −1⟩	⟨0, 0⟩	⟨0, 1⟩	⟨0, 2⟩
⟨1, −2⟩	⟨1, −1⟩	⟨1, 0⟩	⟨1, 1⟩	⟨1, 2⟩
⟨2, −2⟩	⟨2, −1⟩	⟨2, 0⟩	⟨2, 1⟩	⟨2, 2⟩

(··· on left, ··· on right of the ⟨0, ...⟩ row)

$$\vdots$$

Figure 0.5

For each rational can be represented by a pair. (For example, ½ can be represented by the pair ⟨1, 2⟩ and −1½ by the pair ⟨−3, 2⟩.) Yet it seems obvious that there are more rationals than natural numbers since the former include the latter and a lot more besides.

(iii) The paradox of the two men in heaven and hell

Suppose that one man has been in heaven and another in hell for all of past eternity, except that for one day in each year (say Christmas Day) they have swapped positions. Despite our intuition that one of them has spent much longer in heaven than the other, we can, in the same way, pair off the days that one of them has spent in heaven with the days that the other has spent there, and therefore indeed the days that each has spent in heaven with the days that he has spent in hell.

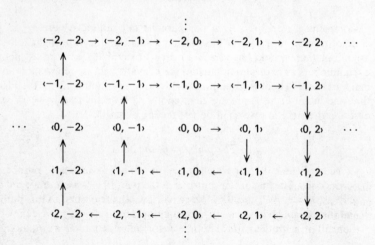

Figure 0.6

Comment: It is clear that many other variations on this theme could be devised, and such variations have long been familiar, as we shall see. This one derives from a suggestion made by Denyer.[5]

(iv) The paradox of the hotel

Suppose there is a hotel with infinitely many rooms, each occupied at a particular time. Then a newcomer can be accommodated without anybody having to move out; for if the person in the first room moves into the second, and the person in the second room moves into the third, and so on *ad infinitum*, this will release the first room for the newcomer. Indeed infinitely many newcomers can be accommodated without anybody having to move out; for if the person in the first room moves into the second, and the person in the second room moves into the fourth, and the person in the third room moves into the sixth, and so on *ad infinitum*, this will release the infinitely many odd-numbered rooms. And if, when all the guests have settled into their new rooms, each is dismayed by how small a bar of soap has been left in the wash-basin, then they can systematically shunt bars of soap along the rooms to ensure that each has two bars instead, or indeed a hundred. All of this puts, to say the least, a strain on our intuitions.

Comment: Hilbert used to present this paradox in his lectures, though some of the embellishments are due to Benardete.[6] The hotel need only occupy a finite amount of space, incidentally. For if each successive floor is half the height of the one below it, then the entire hotel will be only twice the height of the ground floor. This does however raise the problem of what somebody would see who looked at the hotel from above with the roof prized off. (This point is also due to Benardete.[7])

3 Paradoxes of the one and the many

These are paradoxes that pivot on the very idea of considering *one* collection of *many* things, the idea that lies at the heart of set theory and therefore, many would say, at the heart of mathematics – certainly at the heart of contemporary formal work on the infinite. The crispest of these paradoxes are technical paradoxes that arise within set theory, and they require stage-setting that will not be available until the relevant historical background has been filled in (see below, Chapter 8, §§2 and 4, and Chapter 10, §1). But it is already possible to say something about them.

Let us return to the idea of a set. Cantor defined a set as follows:

By a 'set' we mean any gathering into a whole ... of distinct perceptual or mental objects ...

Again:

A set is a many which allows itself to be thought of as a one.[8]

One important consequence of the underlying intuition here is that a set is determined by its members. Typically the members are specified in one of two ways: by citing some condition which they, and they alone, satisfy; or by simple enumeration of them. Thus, for example, we might characterize a set as the set of planets in the solar system. Or we might characterize the very same set as the set whose members are Mercury, Venus, Earth, Mars, Jupiter, Saturn, Uranus, Neptune, and Pluto, which we can write as follows:

{Mercury, Venus, Earth, Mars, Jupiter, Saturn, Uranus, Neptune, Pluto}.

These are two characterizations of the same set, because a set is the set it is solely in virtue of which things belong to it, irrespective of how they have been specified.[9]

The idea of a set is basic and intuitive. This is born out by the fact that Cantor's definitions are hardly more fundamental than what they serve to define. It is therefore particularly alarming to discover that the idea is intimately bound up with certain deep paradoxes. But it is, and they are in many respects the purest of the paradoxes of the infinite.

We can gain a feel for them simply by considering the question: are there any infinite sets? On the one hand we seem bound to say that there are. Take the natural numbers. These are well-defined mathematical entities, forming a totality about which we can make various generalizations. There can surely be no objection to our considering the set of them, and this set must be infinite. On the other hand it seems that for there to be infinitely many things of a given kind is precisely for them to *resist* being collected together in this way. Even the paradoxes of the infinitely big suggest this; for a set is something with a determinate size, but it is precisely when we think of the infinite as having a determinate size that those paradoxes get a grip. Is not an infinite totality a many that is too *big* to count as a one – a many that is ineluctably such?

Although the (semi-technical) idea of a set helps to put the paradoxes of the one and the many into particularly sharp focus, such paradoxes are liable to arise whenever there is a question of trying to recognize unity in infinite diversity. Given the power of the mind to abstract and to unify, it will always look as if this must be possible. Yet at the same time, given the nature of the infinite, it will never look as if it *can* be. Here, perhaps, is an early clue as to why there should have arisen these two conceptions of the infinite, the metaphysical and the mathematical. For it may be that the metaphysical conception is a response to the first of these pulls and the mathematical conception a response to the second. (Hence the sense of

10

conflict between them.) At any rate, the paradoxes of the one and the many, in their different guises, will prove to be a linchpin of the whole enquiry.

4 Paradoxes of thought about the infinite

We turn now to the final group of paradoxes. These are much less clearly delineated than those in the other three groups, but more fundamental. At their root there is a kind of second-order paradox, resting on the backs of all the others so far considered. One radical solution to all of them would be to abandon the concept of the infinite as incoherent. (Without it none of them can properly get off the ground.) So they put collective pressure on us to do that. On the other hand we can feel equally strong pressure from elsewhere to retain the concept. It is true that reflection on the nature of space and time now seems less decisive than it might once have done, because, now that we have greater scientific insight, we are no longer sure that either space or time is infinitely big (infinite by addition) or infinitely divisible (infinite by division).[10] Still, it at least seems to make perfectly good sense, mathematically, to suppose that they are, even if it is false; and this is enough for the concept of the infinite to be coherent. Again, consider the natural numbers: whether or not they can be collected together into a single set, we surely want to be able to say that there are infinitely many of them. But perhaps the strongest pressure to retain the concept of the infinite comes from a rather nebulous, though powerful, sense of our own finitude. This is something which cuts deeper than our awareness that we are mortal and limited in size, constrained in various ways, and ignorant of so much (though it incorporates all of these). It is a sense of being cast into a world that is completely independent of us, most of which confronts us as something alien, something *other* than us, something that impinges on us from without and limits us. (I am not denying that there can be value in overcoming this sense. I shall return to this point at the very end of the book.) This instils in us the idea of a contrast: the idea that the world as a whole – the universe – cannot, in its self-contained totality, be similarly limited by something beyond it, because it includes everything. It must be infinite. One of the paradoxes of thought about the infinite, then, is that there are reasons both for and against admitting the concept of infinity.

A possible solution to this paradox would be to admit the concept of infinity, but to acknowledge (what the earlier paradoxes show) that we cannot do anything with it. That is, we cannot get our minds around the infinite, or discuss it, or define it, or come to know anything about it, or say anything coherent about it. For if we attempt to do any of these things, we automatically abrogate it – because of our own finitude – and become embroiled in contradiction. Any attempt to define the infinite, for example, is an attempt to bring it within our conceptual grasp, but, given our own limitations, we can only bring within our conceptual grasp what is itself suitably limited.

There is something very compelling about this line of thought. But it gives rise to a paradox of its own, perhaps the most serious of all. This paradox is that it seems impossible to reconcile such a line of thought with our having just followed it through. Consider: if we cannot come to know anything about the infinite, then, in particular, we cannot come to know that we cannot come to know anything about the infinite; if we cannot coherently say anything about the infinite, then, in particular, we cannot coherently say that we cannot coherently say anything about the infinite. So if the line of thought above is correct, then it seems that we cannot follow it through and assimilate its conclusion. Yet this is what we appear to have done. We appear to have grasped the infinite as that which is ungraspable. We appear to have recognized the infinite as that which is, by definition, beyond definition. This is the paradox that provides the main focus for this book. It seems to me that a proper reaction to it is a key to the whole enquiry.

So much for paradoxes of the infinite. I now want to say something about the shape of this book. It is divided into two parts. It is in Part One that I outline the history of thought about the infinite.[11] In Part Two I try to address the important issues that arise along the way, including those that have been brought to light in this introduction.

We shall see in Part One that almost all the great philosophers had something important to say about the infinite, and in many cases it was of deep concern to them. Much of what they said was guided by a desire to avoid one or another of the paradoxes outlined above. None of those paradoxes will be very far from the surface at any point in what follows.

It will not have escaped notice that the paradoxes lean to the mathematical side of the topic. And indeed many of those who feature in the history of the topic do so because of the importance of their mathematical work, consisting often of brilliant technical innovations that had repercussions in the very foundations of mathematics. Nevertheless, this book is concerned with all aspects of the infinite, mathematical and non-mathematical alike (as my remarks about the centrality of the fourth kind of paradox ought to have suggested; for paradoxes of thought about the infinite are certainly not – exclusively – mathematical). Metaphysical concepts are to the fore in the book alongside mathematical concepts. For one thing, one of the main tasks that I have said needs to be undertaken is to try to give an account of how these hook up with one another. Insofar as there is an apparently disproportionate emphasis on mathematical issues, it is simply because I take them to provide a particularly clear model of the broader issues. The source of our difficulties with assimilating the mathematically infinite is after all the same as the source of our difficulties with assimilating the infinite more generally: our own finitude.

Our own finitude must be prevalent in any enquiry we conduct into the infinite – if only because, given the paradoxes of thought about the infinite (however they are to be solved), it is clear that we are better able to confront the infinite through analogies and contrasts than head-on. But this is also why the main focus of the book is provided by the last of those paradoxes. For that paradox is itself primarily a matter of the difficulties we have, as finite beings, in trying to assimilate the infinite. What we are seeking then is nothing less than an account of our own finitude, and of our relation to the infinite.

Part One
The History

Philosophers have traditionally concerned themselves with two quite disparate tasks: they have, on the one hand, tried to give an account of the origin and structure of the world and, on the other hand, they have tried to provide a critique of thought. With the concept of the infinite, both tasks are united. Since the time of Anaximander *to apeiron* has been invoked as a basic cosmological principle. And the conceptual change that occurs as *to apeiron* of the Presocratics is refined and criticized by Plato and Aristotle, to the development of Cantor's theory of the transfinite and its critique by Brouwer, is one of the great histories of a critique of reason.

(Jonathan Lear)

CHAPTER 1

Early Greek Thought

It is incumbent upon the person who treats of nature to discuss the infinite and to enquire whether there is such a thing or not, and, if there is, what it is ... [And] all who have touched on this kind of science in a way worth considering have formulated views about the infinite.

(Aristotle)

1 Anaximander and *to apeiron*[1]

The Greek word '*peras*' is usually translated as 'limit' or 'bound'. '*To apeiron*' denotes that which has no *peras*, the unlimited or unbounded: the infinite.

To apeiron made its first significant appearance in early Greek thought with Anaximander of Miletus (c. 610 BC to shortly after 546 BC). Its role was very different from that which it tends to play in modern thought. It was introduced in response to what was then (and has remained) a basic intellectual challenge: to identify the stuff of which all things are made. What, as the Greeks would have put it, is the 'principle' of all things? Thales had earlier proposed that it is water. Perhaps he had been impressed by the natural processes whereby the sea evaporates under the influence of the sun, then forms clouds, dissolves in the form of rain, and soaks into the earth, moistening the food by which living things are nourished. Still, why single out *water* in this way as anything more than just one of the many forms that basic stuff could take? There was something arbitrary about this. So Anaximander's proposal was that the primal substance of which all things are made is *to apeiron*. This he conceived as something neutral, the boundless, imperishable, ultimate source of all that is.

But it was not just that. It was also something divine, something with a deeper significance. Given the processes whereby substances change into one another, the losses and compensating gains, it made good metaphysical sense to suppose that there was an underlying changeless substratum. But for Anaximander it made as much ethical sense. He saw, in

17

the multifarious activity that surrounds us, disharmony and imbalance. He held that opposites were in continual strife with one another (hot with cold, dry with wet, light with dark, . . .); they were continually encroaching on one another (day giving way to night, night giving way to day, . . .) and continually committing injustice against one another. He believed that they must, in time, return to *to apeiron* in order to atone. There they would lose their identity, for where there is no *peras* there are no opposites; and all strife would be overcome.

Anaximander's concerns were at once scientific, philosophical, and ethical. He would not have recognized the modern distinction between empirical hypotheses about the physical nature of the world and *a priori* reflections on how things must or ought to be. We shall see later Greek thinkers to some extent disentangling these strands. Aristotle, in particular, put an empirical gloss on many of these ideas, though he continued to recognize their theological overtones. But Anaximander was simply interested in knowing what the world was like, in the most general sense.

One consequence of this is that it is hard for us to know how seriously to take the materiality of *to apeiron*. It could not be identified with water, or gold, or anything else of such a specific kind: these were at most limited and determinate aspects of it. But could it even be identified with matter? Or was being material and occupying space already a way of being limited and determinate and having a *peras*? If *to apeiron* was *not* material, but something utterly transcendent, then it was signalling what was to become a pervasive and characteristic feature not only of Greek thought but of much subsequent philosophy: the idea of a radical distinction between appearance and reality, where the former includes all that we ever actually come across and the latter is what underlies and makes sense of it. But it is not clear whether such a radical metaphysics was Anaximander's. He talked of *to apeiron* as 'surrounding' us. He may have meant this quite literally.

Given this uncertainty, we find that, when we return to the distinction drawn in the introduction between the metaphysical and the mathematical, we cannot say definitely that Anaximander was working in either territory. For example, given that we cannot even be sure that *to apeiron* was spatial, we certainly cannot be sure that it was mathematically infinite. Later on in this chapter we shall see clear early signs of the polarization between the mathematical and the metaphysical, with concepts of both kinds beginning to filter through into Greek consciousness. But at this early stage in the story we do best not to press the categories. After all, *to apeiron* was radically indeterminate: it was supposed to resist any easy classification.

What does emerge from Anaximander's thinking is a sharp awareness of our own finitude and of the finitude of the ephemera around us, characterized by their generation and decay. I suggested in the introduction how such an awareness might at the same time be an awareness of the infinite. Anaximander, certainly, could make no sense of such finitude except in terms of that which is unlimited and unconditioned; that which suffers neither generation nor decay, so ensuring that the patterns of change that we observe never give out; that into which the ephemeral is destined, ultimately, to be cast back. As he himself put it, in what is the oldest surviving fragment of western philosophy:

> The principle and origin of existing things is *to apeiron*. And into that from which existing things come to be they also pass away according to necessity; for they suffer punishment and make amends to one another for their injustice, in accordance with the ordinance of time.[2]

2 The Pythagoreans

By the time of the Pythagoreans there had been a remarkable turn-about. Pythagoras (born c. 570 BC) was an Ionian. He founded a religious society in Crotona in southern Italy. Central to the outlook of its members was a passionate belief in the essential goodness of what had seemed to Anaximander essentially bad. Where he saw disharmony, imbalance, and strife, they saw harmony, order, and beauty. The regular cycles of the planets, the recurring patterns in nature, the finely proportioned structures in the physical world – these all betokened, for the Pythagoreans, rhyme and reason; that which is comprehensible and good; that which has a *peras*. *To apeiron*, by contrast, was something abhorrent.

It was now unquestionably being conceived also as something spatial. More specifically it was a dark, boundless void beyond the visible heavens. They believed that because it had no end in the sense of limit (*peras*), it equally had no end in the sense of purpose or destiny (*telos*). It was senseless, chaotic, indeterminate, and without structure, simply waiting to have a *peras* imposed upon it. For they quite generally assimilated what has a *peras* to what is good (and to what is one, or odd, or straight, or male, among other things); and they correspondingly assimilated *to apeiron* to what is bad (and to what is many, or even, or curved, or female, among other things). These assimilations were part of a table of opposites that they recognized, whose two fundamental principles, or heads, were *Peras* and *Apeiron*. They believed that the world was the result of an imposition of the former on the latter, a planting of the seed of *Peras* into the void of *Apeiron*. What issued was a beautifully structured, harmonious whole whose parts were held together in unity precisely because of their limitedness and finitude. And the world continued to 'breathe in', and at the same time to subjugate, the surrounding *apeiron*: by doing so, it structured it, and ordered it, and gave it definite shape.[3]

Integral to this picture were the natural numbers (see above, Introduction, §2). These for the Pythagoreans were the key to everything. For it was in their terms, most characteristically in terms of finite numerical ratios, that the imposition of *Peras* on *Apeiron* was to be understood. The Pythagoreans were particularly impressed, for example, by Pythagoras' own (alleged) discovery that musical harmony can be understood in such terms. If the ratio of the lengths of two tuned strings is 2:1, then the shorter sounds an octave higher than the longer; if 3:2, then a fifth; and if 4:3, a fourth. In radical contrast to Anaximander, the Pythagoreans believed that everything that ultimately made sense made such sense as this. (There is a connection here with the very fact that we use the word 'ratio' as we do, and talk of 'rational' numbers (see above, Introduction, §2): 'ratio' in Latin means *reason*.[4])

Natural numbers took on a mystical significance for the Pythagoreans. For example, the first four, those involved in the musical intervals – they discounted 0 – add up to 10, which they held to be a perfect number. This sum is illustrated in the symbol known as the tetractys (see Figure 1.1), a symbol which they believed to be sacred and by which they swore. Indeed not only were natural numbers the key to everything in this scheme, ultimately they *were* everything. They were everything because in the last

Figure 1.1

analysis there was nothing else to which intelligible reference could be made. When *Peras* was imposed on *Apeiron*, numbers were what resulted. The world was a system of structures built within a void, each definable in numerical terms and together constituting a glorious musico-mathematical whole. (This idea, in a suitably modern guise, still has adherents. Mathematics plays a crucial role in the most fundamental scientific theories; and it is still possible to cherish the hope of being able to account for all physical phenomena by appeal to their formal or structural properties, in essentially mathematical terms.)

But the Pythagoreans' veneration of the natural numbers, and their abhorrence of *to apeiron*, were to receive a rude shock.

Pythagoras himself, as is well known, is said to have discovered that the square on the hypotenuse of a right-angled triangle is equal to the sum of the squares on the other two sides. This means, for example, that when the ratio of the hypotenuse to one side is 5:4, then its ratio to the other side is 5:3, as illustrated in Figure 1.2. This is because $4^2 + 3^2 = 16 + 9 = 25 = 5^2$.

Now consider a square (see Figure 1.3). What is the ratio of the diagonal to each side? We can soon calculate, using Pythagoras' theorem, that if

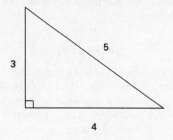

Figure 1.2

each side is 1 unit long, then the diagonal is $\sqrt{2}$ units long (since $1^2 + 1^2 = (\sqrt{2})^2$). But how do we express this as a finite numerical ratio? In other words, which natural numbers p and q are such that the ratio of the diagonal to each side is $p{:}q$?

Effectively what we are asking for is a pair of natural numbers p and q such that $p/q = \sqrt{2}$; that is, $p^2/q^2 = 2$; that is, p^2 is twice q^2. The pair 7 and 5 comes close, but 7^2 differs by 1 from twice 5^2. The pair 17 and 12 comes closer still, but there is still this difference of 1. Indeed it is possible to prove that if the pair x and y misses by 1 in this way, then the pair $x + 2y$ and $x + y$ does likewise. So starting with the pair 7 and 5, or indeed a couple of stages earlier with the pair 1 and 1, we can set up an endless sequence of pairs

⟨1 and 1, 3 and 2, 7 and 5, 17 and 12, 41 and 29, 99 and 70, . . .⟩

each coming closer than its predecessor but each, frustratingly, still missing by 1. Can we then find a pair *not* in this sequence which satisfies our requirement?

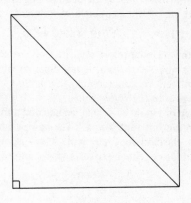

Figure 1.3

The Infinite

The Pythagoreans proved, not much later than 450 B.C., that we cannot. Their proof went as follows.

The Pythagoreans' proof

Suppose that there *is* a pair of natural numbers such that the square of one is twice the square of the other. Then there must be a pair with no common factors (the number 1 does not count as a factor here): for obviously we can, where necessary, divide through. Let p and q be such a pair. Then $p^2 = 2q^2$. This means that p^2 is even, which means, in turn, that p itself is even. So q must be odd, otherwise 2 would be a common factor. [Comment: It is not surprising that the Pythagoreans should have noticed this, given that odd and even occurred in their table of opposites.] But consider: if p is even, then there must be a natural number r such that $p = 2r$. Therefore $p^2 = 4r^2 = 2q^2$. Thus $2r^2 = q^2$, which means that q^2 is even, which means, in turn, that q itself is even, contrary to what was proved above. There cannot after all be a pair of numbers such that the square of one is twice the square of the other.

For the Pythagoreans this was nothing short of catastrophic. The diagonal of a square is incommensurable with each side, showing, apparently, that not everything is to be understood in terms of natural numbers, as they had believed. $\sqrt{2}$ is not a 'rational' number.

They had various ways of trying to cope with the catastrophe (though according to legend, one of them was shipwrecked at sea for revealing the discovery to their enemies). The point, however, is that it presented them, an unsuspecting, unreceptive, and unwilling audience, with a first real glimpse into the mathematically infinite. The natural numbers, each only finite, had proved unequal to an apparently simple mathematical task. Something transfinite now had to be acknowledged, for example something lying beyond the infinite sequence of rationals

$$\langle 1/1,\ 3/2,\ 7/5,\ 17/12,\ 41/29,\ 99/70,\ \ldots \rangle$$

Whether or not the Pythagoreans regarded *to apeiron* itself as mathematically infinite is not clear, for it is not clear to what extent they had consciously assimilated the mathematically infinite. But here it was, effectively showing up in their very midst.

One consequence of all of this is that we can add a third kind of number to those defined in the Introduction, §2. The new numbers can be thought of – (this will be clarified in Chapter 4, §3) – as those which can be expressed using infinite decimal expansions. Examples are:

$$2\ (=2.000\ldots);$$
$$\tfrac{1}{3}\ (=0.333\ldots);$$
$$\pi\ (=3.141\ldots);$$

22

and, of course,

$$\sqrt{2} \ (= 1.414\ldots).$$

Some, but as we have seen not all, of these numbers are also rationals. (π, incidentally, is not – though its irrationality was not established until the eighteenth century, by Lambert.) Any that is *not* has a decimal expansion which is not susceptible of finite abbreviation; that is, it does not consist, after a certain point, of some recurring finite pattern. To add insult to the Pythagorean injury, such numbers, rationals and irrationals alike, are known as the *real numbers* (or *reals*).[5]

3 The Eleatics

Parmenides and Melissus

Parmenides of Elea (born c. 515 BC) was a Pythagorean who rebelled. He rebelled not because of $\sqrt{2}$ but because he was dissatisfied with the Pythagorean conception of the world as a system of structures within a void. As an ultimate explanation of everything that is, this left Parmenides dissatisfied because it seemed to involve ineliminable reference to what is not. (The void, in *some* sense, is not.) For Parmenides this was incoherent. He believed that reality – The One – must be autonomous and explicable in its own terms, a perfect unified self-subsistent whole. In other words, in the language of the Introduction, though not, as we shall see, in Parmenides' own language, he believed that reality must be metaphysically infinite. This is a significant point in the history of the infinite. It is here, for the first time, that metaphysical concepts put in a recognizable appearance – though they have not yet been explicitly related to the infinite.

Parmenides' rebellion took the form of something like a return to the doctrines of Anaximander, but under the most radical of interpretations. Reality, on Parmenides' view, had to suffer no change, because change involves a transition from what is to what is not and, as we have seen, there could not be any ultimate appeal to what is not. In particular, reality had to suffer neither generation nor decay, either as a whole or in its parts. Indeed there was no sense in talking of its parts. For how were they to be distinguished without appeal to the void – without appeal to those contrasts that can only be supplied by what is not? Reality was indivisible, homogeneous, and eternal. (It was eternal in the sense of being timeless; for how could there be time if there could be no change?)

As a result of these views Parmenides was forced to accept a radical distinction between reality and appearance of the kind already alluded to in connection with Anaximander. All that we actually encounter, in its transitoriness and diversity (all that the Pythagoreans had held so dear) was deemed an illusion. It was how reality appears to us. 'The Way of

Truth' led us to reccognize how things must really be: 'The Way of Opinion' concerned the multifarious, mutable way they seem.[6]

Parmenides' views about reality involved the clearest possible expression of a conception of the metaphysically infinite, and they were close in many ways to the views of Anaximander. Nevertheless, he did not himself speak of reality as *to apeiron*. On the contrary, he likened it to a finite sphere. He wrote:

> Powerful necessity holds it enchained in a limit which hems it around . . .

and again:

> Since there is a last limit, it is completed on all sides, equal in every way from the middle, like the mass of a well-rounded ball.[7]

The metaphysical conception of the infinite had not yet come to the forefront of Greek thinking.

It was there, however, in the background. We must not forget that Parmenides was only offering an analogy. He was intending to convey the unity of reality, through the idea that it could be regarded equally from every point of view. (Here it is interesting to note that an earlier thinker, Xenophanes, had espoused views similar to Parmenides'; and whereas he had expressly argued that we could not intelligibly describe reality either as infinite or as finite, he too had held reality to be the same from every point of view, in such a way that later commentators ascribed to him the view that it was a finite sphere.[8]) If it is true that Parmenides had effectively, if unwittingly, embraced the metaphysically infinite, then it in fact makes perfectly good sense that he should have found it natural to invoke an analogy with what is finite. Limits are not always imposed on something from without by something else; they can be imposed on it from within by its very own nature, as in this case. Once we admit this, then the metaphysically infinite, like the mathematically *finite*, is bound to be limited.[9]

This is one of the reasons why Parmenides would not have wanted to use the term '*apeiron*' to describe reality. He was sensitive to its increasingly mathematical overtones. The Greeks were still not sure what sense to make of these. But they certainly conveyed a sense of the incomplete (of that only a part of which is ever present), exactly *not* the terms in which Parmenides wanted us to think of The One. (His amplification of the first of the two quotations above was: '. . . because it is right that what is should be not incomplete.') Here, in effect, are the first signs of tension between the two conceptions of the infinite, the metaphysical and the mathematical.

Parmenides founded what became known as the Eleatic school, after his home town. And it took one of the members of this school, Melissus of Samos (lived fifth century BC), at last to venture a metaphysical understanding of infinitude and to declare The One to be infinite. (He did not think that it was *mathematically* infinite. He expressly denied, for example, that it was extended, lest it should have parts.) He was not disagreeing with Parmenides. On the contrary, he was putting the Parmenidean view in a particularly succinct way.[10]

Zeno

Another member of the Eleatic school was Zeno of Elea (probably born c. 490 BC). Zeno is famous above all for his four paradoxes of motion, of which the best known is the paradox of Achilles and the tortoise – the first of the paradoxes of the infinitely small that I outlined in the introduction. Let us look at the remaining three, themselves paradoxes of the infinitely small. Two can be recast as follows.[11]

(i) The paradox of the runner

Suppose that Achilles wants to run straight from A to B. First he must run to the mid point between them. Then he must run to the three-quarter point. And so on *ad infinitum*. It seems that Achilles can never arrive at B, which is absurd.

Comment: This is essentially the same as the paradox of Achilles and the tortoise. (Think of A as Achilles' starting point in the race and B the point at which he will overtake the tortoise.) It was later sharpened by Aristotle, who envisaged someone running from A to B and counting each time he passed one of the assigned points: on arrival at B he would have to have counted infinitely many numbers, which seems impossible.[12]

(ii) The paradox of the arrow

Whatever occupies its own space throughout a period of time is at rest throughout that period. So at any instant an arrow (say) must be at rest. But this is tanatamount to saying that the arrow cannot move, which is absurd.

The fourth paradox is less clear. It concerned three bodies A, B, and C, the last two of which were moving in opposite directions at the same speed relative to the first. It may have been directed against a discrete conception of space and time, for it related to the fact that B was moving twice as fast relative to C as to A: on a discrete conception, it is possible to convince oneself that all motion (even relative motion) must be to the next point at the next moment and therefore at the same speed.[13]

However this final paradox was to be taken, they collectively made an enormous impact on the history of the infinite, and presented a lasting challenge, as we shall see.

Not that Zeno himself intended any of them as paradoxes though. He was trying to defend Parmenides' views by showing the unreality and incoherence of change (specifically of motion). Some of his arguments told against a discrete conception of change, others against a continuous conception. Their overall message was that no sense could be made of change on any conception. At the same time, of course, they reinforced the growing suspicion that no sense could be made of the mathematically infinite either. In one of Zeno's further arguments in defence of Parmenides, this suspicion became an outright assumption. The argument was designed to show that reality must be a unity and not a plurality, a one and not a many. It ran as follows.

> *Zeno's argument*: If reality were a many (alternatively, if reality were how it appears to be – having parts between any two of which there is a third), then it would have to have infinitely many parts. But there cannot be infinitely many of anything. So reality must be a one.[14]

Comment: If we think that reality does have infinitely many parts, yet share Zeno's qualms about the infinite, then we can regard this as a particularly raw example of a paradox of the one and the many.

Zeno may also have formulated a version of the paradox of the divided stick. It is anyway clear that the mathematically infinite, like the metaphysically infinite, was now finally impinging on Greek consciousness; and that hostility towards it, at least in some quarters, had already become fierce.[15]

4 Plato

Plato (c. 428 BC–347 BC) was an Athenian and is generally acknowledged to have been one of the most brilliant thinkers of all time. We get a sense of his genius from his handling of these issues. He approached *to apeiron* in the same way that he approached so many other topics, managing to achieve a remarkable synthesis of what had been outstanding in the views of his predecessors with his own original insights – thereby displaying eclecticism of the best kind (such as we shall not see repeated until Kant). In line with Anaximander he recognized the problem of conflict between opposites. His solution, however, was Pythagorean. And the attendant metaphysics was in many respects Eleatic.

This gloss is subject to an important *caveat* though. Plato was a dialectician. He probed ideas, toyed with them, teased out their consequences. One thing that enabled him to do this was the fact that he wrote in dialogue form (and nearly always adopted Socrates, with his zetetic methods, as protagonist). We must therefore be wary of attributing definite views to him. For example, from one of his dialogues, the *Timaeus* (which was admittedly more of a monologue), there emerged a remarkably modern and beautiful atomistic account of how the four newly acknowledged elements – fire, air, earth, and water – were able to interact and change into one another. Fire was composed of minute particles shaped as tetrahedra; earth, of cubes; air, of octahedra; and water, of icosahedra. These particles were themselves composed of triangles. The particles could be broken apart into their constituent triangles, and these could be rearranged to effect the various macroscopic changes that we observe. Here already was a well worked out and very Pythagorean approach to what had been one of the basic problems to exercise Anaximander. Yet in another of his dialogues, the *Parmenides*, probably written later, he projected arguments for the infinite divisibility of matter that seemed to tell against this doctrine.[16] We are unlikely to get the best out of Plato, then, unless we approach his dialogues in the same exploratory frame of mind in which, it seems, they were written.

A system can none the less be discerned in them, and it contains a distinctive account of *to apeiron*. Anaximander had been right to focus on opposites. But it was not that these arose from *to apeiron* and needed to return to *to apeiron* in order to atone. They already constituted *to apeiron*. For Plato's conception of *to apeiron* was much more abstract than either Anaximander's or the Pythagoreans' had been. It was whatever admits of degrees and contains opposites within its range. A good example of this was temperature, containing within its range hot and cold. *To apeiron* was the indeterminate.

But the qualities of things round about us are quite determinate. My body temperature, for example, is 98.4°F. What was Plato's account of this?

It was just as the Pythagoreans had maintained, the effect of a general imposition of the *peras* on *to apeiron*. This effect was always to produce some particular (numerical) value within the given range – a particular temperature in this example. It was as if *to apeiron* determined what the possibilities were, and the *peras* was imposed on it to determine which of them was to be realized. (The atomistic theory outlined above might be seen as the minute workings out of this.) But this need not result in any strife or injustice. Conflict between opposites could be resolved by the *peras* holding them in a harmonious balance. Musical harmony, for example, was the result of combining notes of particular pitch. Of course, they had to be the *right* pitch (just as my health is a result of my body being at the *right* temperature). We shall see shortly what it was, on Plato's view, that ensured the right values, in other words what controlled how the *peras* was to be imposed on *to apeiron*. But the effect was a world that was, as the Pythagoreans had held it to be, a beautifully ordered whole.

This was not the real world however. It was the sensible world, to which we have empirical access – a world of appearances. The real world was transcendent. Plato accepted a distinction between reality and appearance that was every bit as radical as Parmenides'. And what controlled how the *peras* was to be imposed on *to apeiron* was some kind of cosmic cause, a principle of intelligence or reason, that belonged to the real world.[17]

Plato believed that everything that was ultimately good belonged to the real world. It was a world of Ideas (one might also say Ideals). These were archetypes of things in the sensible world, which imperfectly imitated, or 'participated in', them. For example, particular acts of justice participated in the Idea of justice. These Ideas were eternal and immutable. Supreme among them was the Idea of the good. It was in terms of this that everything else was to be understood. Of the sensible world we could have only opinion. True knowledge (for example, our knowledge of what justice is – our knowledge of the Idea of justice) was knowledge of what was real.[18]

The unity of the Ideas, encapsulated in the supremacy of the Idea of the good, meant that there was an element of the metaphysically infinite in reality. But Plato no more spoke in these terms than Parmenides had done. It was not just that he had his own account of *to apeiron*. There was also a lingering Pythagorean resistance to the idea that there could be anything infinite about what was real and true and good. Admittedly Plato held Ideas to be eternal. (This meant that our epistemic access to them had to be viewed as a kind of eternal knowledge, which in turn fuelled his theory of immortality.[19]) But, at least in this context, he meant eternity in the same way as Parmenides, as timelessness.[20] Any legitimate concept of the infinite, other than his own somewhat idiosyncratic concept of *to apeiron*, would have to have its home, it seemed, in the world of appearances.

To an extent, it did. Not that Plato held this world to be spatially infinite. On the contrary, he argued that it was spherical, appropriating arguments similar to those applied by Parmenides to the real world.[21] But he did recognize in it infinite diversity.[22] (This was one reason why the relationship between an Idea and the things that participated in it presented a particularly acute paradox of the one and the many. How was unity to be recognized in such infinite diversity? How, for example, could infinitely diverse acts all count as acts of justice? Was it enough to say that they participated in the Idea of justice? Or was there then a further problem of how to recognize unity in the infinitely diverse ways in which they did so, and so on *ad infinitum*? Plato was greatly exercised by perplexities of this kind.[23])

The upshot of all of this, given what Plato took to be ultimately real and given where his interests ultimately lay, was that he had very little concern for the infinite in its – now increasingly familiar – mathematical guise. There was no real incentive for him, any more than there had been for so many of his predecessors, to engage with the thorny issues that surrounded it.[24]

5 Early Greek mathematics[25]

What then of early Greek mathematics?

It would be a mistake to think that there were no Greeks who were prepared to look the mathematically infinite squarely in the face. Archytas

28

of Tarentum (flourished 400 BC–365 BC), who was a Pythagorean and a friend of Plato, presented what now strikes us as a most primitive and natural argument for the spatial infinitude of the universe. In a version that perhaps owes something to the embellishment of later thinkers, it ran essentially as follows.

> *Archytas' argument*: If the universe had an edge, then we could imagine someone, at the edge, trying to stretch out their hand. Success would show that there was at least empty space beyond; failure, that there was something preventing them. Either way, this would not, after all, be an edge. So the universe must be of infinite spatial extent.

Comment: Spatial infinitude, unlike temporal infinitude, and unlike the infinitude of the natural numbers, seemed, and still seems, particularly serious because of a vague sense of its being there 'all at once' – and an equally vague sense that this matters. These themes will be taken up in the next chapter. For an objection to Archytas' argument, see below, Chapter 9, §2.

Despite the simplicity of Archytas' argument, and despite the fact that √2 presented the Greeks with an early challenge that occasioned further work on irrationals, the infinite was never, as such, an important object of mathematical study for them. Rather, Greek mathematics embraced the infinite in an indirect way, a way that has become an important model for subsequent mathematics. It will also prove to be an important model for this enquiry. We find a perfect example of what I have in mind if we turn to Greek geometry. This did in fact presuppose infinite space, both by addition and by division. For example, any line was taken to be indefinitely extensible and infinitely divisible. Yet, as Aristotle pointed out,[26] there was no explicit reference to infinite space or to anything else infinite: the objects of study in geometry were always finite lines divided at (at most) finitely many points. Arithmetic, which in its technical sense concerns the natural numbers, supplies an even clearer example. It presupposes the existence of infinitely many natural numbers, since each natural number is taken to have a successor (a natural number that is one greater than it). But no natural number is *itself* infinite: the objects of arithmetical study are all finite. The point is, study of what is itself finite is sometimes possible only in an infinite framework.

Euclid (flourished c. 300 BC), who lived at Alexandria, is famous for having axiomatized Greek geometry. That is, he devised a small stock of axioms or postulates, which were taken to be incontrovertible, and from which all the rest of the geometry could be derived.[27] Apart from the fact that his system, as set down in his *Elements*,[28] encapsulated in a particularly stark way the power, rigour, and beauty of early Greek mathematics, it was significant also in demonstrating once again the allure of the finite. For one of the appeals of axiomatizations is surely that they purport to trap an infinite wealth of information or wisdom in a finite, manageable stock of basic (self-evident) principles. (The appeal shows up in various places, for example in St Mark, XII, 28–34; in Spinoza's *Ethics* and Newton's *Principia*, both couched in an axiomatic form; and in the American Declaration of Independence.) The whole issue of axiomatization will greatly occupy us later (see below, Chapter 8, §6, and Chapter 12).

But perhaps the greatest example of how Greek mathematical genius was able shrewdly to circumvent the mathematically infinite was in the work of Eudoxus (c. 408 BC–c. 355 BC), who founded a school in Cyzacus, and Archimedes (c. 287 BC–212 BC), who lived most of his life at Syracuse. Eudoxus established what is known as the method of exhaustion, and Archimedes subsequently much exploited it. This was a method of discovering the properties of curved figures by investigating the properties of polygons acting as successively better approximations to them. For example, Archimedes used the method to find the area of a circle C with radius r. The following is a perversion of his argument, but it is a heuristically useful starting point.

Let C be a circle with radius r. For each natural number $n > 2$, let P_n be a regular n-sided polygon (a polygon with n equal sides and n equal angles) inscribed inside C. P_n can be divided into n congruent triangles, as illustrated in Figure 1.4 for the cases $n = 4$, $n = 6$ and $n = 8$. Let the base of each triangle be b_n and its height h_n (see Figure 1.5). Then the area of each triangle is $\frac{1}{2}b_n h_n$. Thus the area of P_n as a whole is $n\frac{1}{2}b_n h_n$, or $\frac{1}{2}nb_n h_n$. But C itself can be regarded as a polygon with infinitely many infinitely small sides. In other words, C is what we get when we extend the original definition of P_n and allow n to be infinite. When n is infinite, nb_n equals the circumference of $C = 2\pi r$ (where this follows from the definition of π) and h_n equals the radius of $C = r$. So the area of C is $\frac{1}{2}.2\pi r.r = \pi r^2$.

This 'reconstruction' of Archimedes' argument has some intuitive appeal, but it is not ultimately satisfactory and it would not have satisfied Archimedes. We cannot uncritically plug the infinite into equations such as these. (How is multiplying by an infinitely small quantity different from multiplying by 0, for example?) Nor is it legitimate to talk about a polygon with infinitely many infinitely small sides. Or at least, it is not legitimate until clear sense is explicitly conferred on it, and for this it does not suffice to think of the infinite as something resembling a natural number, only bigger.

Of course, part of what is going on here is that, the larger n is, the more nearly P_n approximates to C. But there is more to it than that. It is also true that, the larger n is, the more nearly P_n approximates to the 'deformed' circle C^* in Figure 1.6 (just as the rationals in the sequence

30

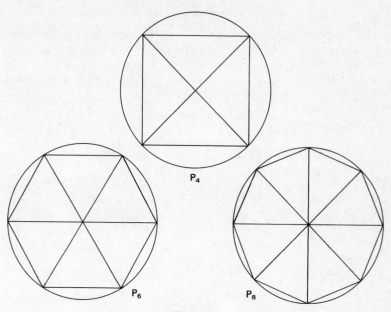

Figure 1.4

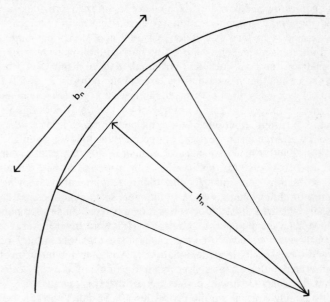

Figure 1.5

31

⟨½, ¾, ⅞, . . .⟩,

as well as getting closer and closer to 1, get closer and closer to, say, 1⅓). The key point, intuitively, is that C, unlike its deformed counterpart C^*, is the *limit* of the polygons; it is what they are 'tending towards'. But it is very hard to see any way of capturing this intuition without, once again, thinking of C as an 'infinigon'.

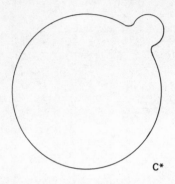

C^*

Figure 1.6

The genius of Eudoxus and Archimedes lay in their providing a way. Displaying a kind of sophistication and rigour that had to be recaptured two thousand years later (see below, Chapter 4, §2), they pinpointed the crucial difference between C and C^* by proving the following: no matter how small an area ε you consider (a trillionth of the area of C, say), there is always some number n such that the area of the polygon P_n differs from that of C by even less than ε. (That is, there is a polygon as close in area to C as you care to specify. This is not true of C^*.) Not only does this eschew appeal to the infinite, but one of the principles on which it rests, which we now know as Archimedes' axiom, can be recast as follows: given any two quantities A and B such that A is greater than B, there is a natural number n such that if A is halved, and the half is halved, and so on n times, this yields a quantity less than B; and this is tantamount to denying the existence of either infinitely large or infinitely small quantities.

Because the area of each polygon is provably less than πr^2, Archimedes concluded that the area of C is at most πr^2. (If it were any bigger, say $\pi r^2 + \delta$, then there would have to be a polygon whose area was within δ of this area and therefore itself greater than πr^2.) A similar argument involving circumscribed polygons establishes that the area of C is at *least* πr^2. Archimedes concluded, with perfect rigour, that it is exactly πr^2.

Painstakingly working on the properties of a 96-sided polygon, he also went on to show that $3^{10}/_{71} < \pi < 3^{1}/_{7}$. Elsewhere he did interesting work exploring economical ways of expressing larger and larger natural numbers, as part of an attempt to say how many grains of sand it would take to fill the known universe. These concerns were very much with the finite. Along with many other aspects of early Greek mathematics they showed that, even at a technical level, there is an element of truth in the common, though admittedly oversimplified, adage: the Greeks abhorred the (mathematically) infinite.[29]

CHAPTER 2

Aristotle[1]

In general, the infinite exists through one thing being taken after another, what is taken being always finite, but ever other and other.

(Aristotle)

1 Preliminaries

Aristotle (384 BC–322 BC) is a touch-stone for this whole enquiry. Born in Stagira, he lived most of his life in Athens, where he studied under Plato in the Academy that he (Plato) had founded. He was a remarkable polymath. He made major contributions to logic, metaphysics, the natural sciences (above all biology), psychology, ethics, politics, and literary criticism; and some of these disciplines he can even be said to have founded. His entry into our particular drama in many respects marks the end of the prologue and serves to inaugurate the action proper. Many of the concepts that have shaped and informed subsequent discussion, indeed much of what has actually been discussed, originated with Aristotle.

He himself began with the views of his predecessors. He noted in particular one recurring and dominant theme in what they had been saying: whatever is infinite is *ipso facto* a 'principle', that is to say something fundamental from which other things are derived or in terms of which other things are to be explained. Otherwise, it would be derivative and thus limited. This is why, despite profound differences in their views, earlier thinkers had all held the infinite to be ungenerable, indestructible, and eternal. It is also why they had tended to refrain from attributing any particular or determinate qualities to it. These too would have counted as limiting it, leaving it open to explanation in more fundamental terms.

Such remarks apply both to those thinkers who regarded the infinite as an entity in its own right (paradigmatically Anaximander, for whom it was actually a substance) and to those who thought of infinitude in a more modern vein as a property that other entities possessed. Among the latter were some philosophers not discussed above. Anaxagoras, for example, had held that all the different substances in the world were originally bound together in a single undifferentiated mass. And this he referred to not as *to apeiron* (*the* infinite) but as something that counted infinitude among its properties. Again, Leucippus and Democritus had formulated a very modern atomistic view, such as Plato was later to set forth (see above, Chapter 1, §4), according to which the world consisted of indivisible atoms circulating within a void. They likewise had taken infinitude to be one of the properties of these atoms. It is worth pausing to reflect on this. Here, for the first time, we see infinitude being predicated of a plurality rather than a single subject. As the idea of a single primordial substance, such as Anaximander's, had given way to the idea of a plurality of primordial substances, so too the locution 'It is infinite' had given way to the locution 'They are infinite'. The point was, of course, that there were supposed to be *infinitely many* of them. The infinite was now figuring in answer to the question, 'How many?' That is, the idea of infinite number, as opposed merely to infinite magnitude, had now emerged. But in all this diversity of thought Aristotle recognized consensus on one point: whatever is infinite must be basic in the scheme of things.

His first concern, then, before he went on to decide how much and what exactly he was prepared to accept, was to make what sense of this he could. And here he was swayed by a certain empiricism.[2] It is not that he was unsympathetic to the idea of anything transcendent. But he did reject the kind of appearance/reality distinction that had found fruition in the Eleatics and later in Plato. The transcendent, on his own conception, enjoyed a much more intimate association with what can be directly experienced. And this meant that for Aristotle, if the infinite was going to play anything like the role(s) that had so far been assigned to it – if it was going to be basic in the scheme of things in anything like the ways envisaged – then sense could only be made of it in essentially spatio-temporal terms. For example, he did not consider the very radical interpretation of Anaximander's views according to which *to apeiron* was immaterial. As far as Aristotle was concerned, the only serious question that had to be addressed was whether anything *in nature* (anything in the world of space and time) was infinite.

He therefore defined the infinite in a somewhat new way, a way that would accord with just such a concern. He defined it not as that which has no limit or bound, but as that which is untraversable. But this was ambiguous, as he realized. Not all of its interpretations lent themselves to his naturalistic conception. Some went precisely against it. For example, something can be said to be untraversable because it makes no sense to speak of traversing it (rather as a voice might be said to be invisible). In this sense, something utterly beyond our experience could count as untraversable. This was not what Aristotle intended. So what did he intend? Even when we say that something physical is untraversable, we can still mean different things, as Aristotle pointed out. We can mean that there is nothing that counts as completing its traversal (as with a uniform circular racecourse); or that it is simply difficult or impossible in practice to traverse it (as with a treacherous river); or, finally, that it goes on for ever, and in that sense has no traversal (in other words, in our terminology, it is mathematically infinite). Aristotle essentially intended the last of these.

This is highly significant. Here, arguably, was the first explicit character-ization of the mathematically infinite – the point at which it was first clearly registered in Greek consciousness. It is not that there had been no earlier thinkers who had referred to, or alluded to, the mathematically infinite. Zeno is an obvious example (see above, Chapter 1, §3); and Anaxagoras had said that, however small anything was, there was something else still smaller, and however big anything was, there was something else still bigger.[3] But nobody else, arguably, had singled the mathematically infinite out and got it into quite such sharp focus as Aristotle. His naturalism had provided him with the perfect forum to do this. Indeed, as we shall see, he was at pains to bring the mathematical conception of the infinite to the fore as the only true conception. He rejected earlier metaphysical tendencies as being fundamentally awry.

He also drew a distinction within the infinite between that which is infinite by addition and that which is infinite by division.[4] (One would speak of traversal, in the case of division, when successive divisions were being effected. The infinite by division is that which cannot be traversed in this sense.) This distinction he attributed to Plato.[5]

Having thus clarified how he understood the question, Aristotle pro-ceeded to enquire whether anything in nature was infinite.

2 The problem

I suggested in the last chapter how far other Greek thinkers abhorred the mathematically infinite. It seemed, from the way in which he went on to address his question, that Aristotle did too. He produced a whole battery of arguments in support of a negative answer.

He first discounted what he took to be Anaximander's view: that the infinite was to be regarded as a material substance in its own right, 'stuff' of the most basic kind. If it were that, it would have to have parts that were in turn infinite (just as air has parts which are air, and water has parts which are water). But Aristotle thought this absurd. Infinitude, on his understanding, had to be a property of the whole – and so, indeed, a *property*. (This already sets the seeds for paradoxes of the infinitely big, as Aristotle was in effect aware. These paradoxes arise precisely from the fact that what is infinite *can* have parts which are infinite, and so, it seems, as great as the whole.)

Aristotle also dismissed the view of the atomists, that infinitude could be a property of a plurality. For like Zeno, he rejected as incoherent the idea of infinite number. A number, he held, is what can be arrived at by counting; and counting to infinity would involve traversing an infinite sequence of numbers. So he did not bother to enquire whether there were infinitely many things of any particular kind.

The only real question, then, for Aristotle, was whether any single thing, for example one of the four recognized elements, was infinite. But this question was in danger of being dismissed equally quickly. For in effect it was a question about the possibility of infinite body. But on one natural understanding, a body is something bounded by a surface and so, by definition, finite. Aristotle was keen not to settle the issue by *fiat*, however. He urged a relaxed understanding of body.

His reasons for rejecting (even so) the possibility of infinite body were largely empirical. Or at least, this is true of his reasons for rejecting the possibility of body infinite by addition. He ruled out the possibility of an infinitely divisible body on essentially conceptual grounds, appealing to the (relevant) paradoxes of the infinitely small. These paradoxes (including Zeno's) revealed, he thought, incoherence in the idea of anything physical's ever being divided into infinitely many parts.

Turning to the infinitely big, he first rejected the idea that something neutral, in the sense of being more fundamental than any of the recognized elements, could be such. This was not only for the reasons given in his rejection of Anaximander's position but also because experience showed, quite simply, that there was no such stuff: experience showed that it was impossible to break things down any further than into air, fire, earth, and water. (Of course, these arguments have subsequently lost much of their empirical underpinning.) On the other hand, he argued, at most one of *these* could be infinite; for it would have to be infinite in every direction, allowing room only for finite portions of each of the others. Yet if any one of them *were* infinite, the others would have been destroyed, in the way that fire, for example, is quenched by a sufficient amount of water. Furthermore, whatever was infinite would have to have a natural 'direction', as earth, for example, tends towards the centre of the universe and fire away from it: but Aristotle could not see how to make sense of this except in finite terms (for example, when it came to identifying the centre of the universe). In any case, if an infinite body had some 'direction', then it would also have to be capable of movement. But Aristotle had a plethora of arguments designed to show that this was impossible. (Some of these were empirical, some conceptual. All would now be regarded as invalid, though in some cases for very subtle reasons. He attempted to show, for example, that an infinite body could move only if there could be movement over an infinite distance in a finite time. But this is not so: any distances traversed, either by the whole or by any of its parts, need only be finite.)

These were all more or less specific considerations. But also, more generally, the very idea of the infinite seemed somehow incoherent: witness Zeno's paradoxes and the like. (Indeed, would not anything infinite have to have infinitely many parts? But Aristotle had denied that there could be infinitely many of anything.) The upshot of his discussion seemed clear enough: nothing could be infinite.

But Aristotle knew only too well that it would not do to let the matter rest there. There were also powerful considerations in favour of recognizing the existence of the infinite, and they had to be addressed. He singled out what he took to be the five most powerful.

(i) Time seemed to be infinite, both by addition and by division.

Comment: The discussion so far had mostly concerned spatial extension, so there *might* not be any deep conflict here.

(ii) Matter seemed to be infinitely divisible; that is, it seemed not to have any ultimate indivisible constituents.

Comment: This meant that the paradoxes of the infinitely small really were *paradoxes*. It was not good enough to present them as part of an argument *against* infinite divisibility, and leave it at that.

(iii) The continual generation and destruction of new things, which was integral to nature as it was then conceived, seemed impossible without an infinite supply of matter.

Comment: This was a consideration that had impressed Anaximander (see above, Chapter 1, §1).

(iv) It seemed that whatever was finite, or limited, was limited by something else beyond, so that there could not be any ultimate limits.

(v) Above all, it seemed to be an *a priori* truth enshrined in mathematics not only that the sequence of natural numbers was infinite but that space itself was infinite.

Comment: I already remarked in Chapter 1 that there was something indirect about the way in which Greek mathematics was involved with the infinite. But indirect involvement is involvement. In any case, the infinitude of space had also been argued for much more directly and in equally *a priori* terms by Archytas (see above, Chapter 1, §5). For Aristotle this issue was particularly urgent. He could not reconcile the infinitude of space, at least by addition, with what he had already assiduously argued. For he believed that if space were infinite, then body would have to be infinite too. This was for two – distinctively Aristotelian – reasons. Firstly, if space were infinite but body only finite, there could be no reason why any particular part of space, as opposed to any other, should house it. Secondly, it at least had to be possible for space to be fully occupied, but, where eternal things like space and body are concerned, whatever is possible must be.

The net effect of all of this was that Aristotle was faced with a dilemma. What he took to be decisive arguments against the existence of anything infinite had to be reconciled with compelling arguments in its favour. Two of the latter he felt he could simply rebut. He rebutted (iii) by appeal to recycling; and he rebutted (iv) by just denying that being limited, unlike, say, being touched, meant being limited by something else (rather as Parmenides had spoken of the limits of reality without thinking that there was anything beyond). But three considerations remained; and so did the dilemma. His problem was a version of what I presented in the introduction as the first of the paradoxes of thought about the infinite. His solution was masterful.[6]

3 The solution: the potential infinite and the actual infinite

We get a clue as to the nature of this solution if we think about the difference between space and time. (My aim in this section is to present the substance of the solution. I shall defer to the next section discussion of how Aristotle actually applied it.) Aristotle was convinced that the world was spatially finite by addition, for the reasons which he had given. As a result he believed the same of space itself. But he was equally convinced that time was infinite by addition. For he believed that any point in time could be recognized as a 'now', and that it was in the nature of a 'now' to divide past from future: no point could be first, and none could be last.[7] He formulated his solution to the problem he faced in terms that paralleled this difference of attitude to space and time. His proposal was essentially this: there is no objection to something's being infinite *provided that its infinitude is not there 'all at once'*. (In Chapter 1, §5, I commented on the intuitive if vague appeal of this idea.) The way Aristotle himself put it was as follows: the infinite exists *potentially* but not *actually*. Slightly differently: something can be potentially, but not actually, infinite.

This was without doubt his greatest legacy to later thought about the infinite. This distinction proved to be enormously influential. But how exactly are we to understand it?

A distinction between the potential and the actual played a key role in Aristotle's thought, quite generally. Something may be actually a piece of wood, for example, but potentially a box. We do best not to think in these terms, however, as Aristotle himself warned. For on this general conception, if something is potentially thus and so, then it must be possible for it to be actually thus and so. If something is potentially infinite, on the other hand, then it is not even possible for it to be actually infinite. The distinction between the potential and the actual infinite was for him more or less *sui generis*.[8]

There have been many subsequent attempts to elucidate the distinction, but we can do no better than to turn back to Aristotle's own elucidations. This means continuing to think of the distinction in essentially temporal terms. The actual infinite is that whose infinitude exists, or is given, at some point *in* time. The potential infinite is that whose infinitude exists, or is given, *over* time; it is never wholly present. (I use the phrase 'is given' advisedly. The metaphor of reception has often been felt to go naturally with this account, since reception takes place in and over time.)

Imagine a clock, for example, endlessly ticking. Its ticking is potentially, but never actually, infinite. It is as if it is in a constant state of becoming but never actually *is*, in its entirety. It never achieves full being. Spread over time, it exists in the same way that a day exists. But, in contrast to the case of a day, there is no time at which it has completely run its course or by which it can be said to have been actualized. True, the successive tickings of the clock are a kind of actualization of its overall ticking, but one which is continual and never-ending, not such as to threaten the view that what is potentially infinite must always fall short of being actually so. Many philosophers since Aristotle have tried to see this appeal to time as an attempt to grasp, by means of a metaphor, something deeper, more abstract, and more significant: but Aristotle himself took it quite literally.

Nor was his doing so unmotivated. For Aristotle the infinite was the untraversable. But traversal takes time. So there is no making sense of the claim that something is untraversable save with respect to the whole of time. Furthermore, Aristotle believed that questions of possibility and impossibility were themselves intimately connected with time, so that asking whether or not something was possible was akin to asking whether or not it would be so – at some time.[9] In particular, asking whether or not something was untraversable was akin to asking whether or not it would ever be traversed; and to say that it would not was thus a way of making a generalization about the whole of time. It transpires, then, that the idea of the actual infinite – of that whose infinitude presents itself all at once – was close to a contradiction in terms for Aristotle. By contrast, the claim that time was infinite by addition, properly understood as meaning that it was potentially infinite, was close to a tautology.

4 Application of the solution

The solution thus set up, Aristotle proceeded to apply it. His battery of (largely empirical) arguments against the existence of the spatially infinite by addition he could let stand. What he now needed to show was basically that, where it seemed necessary to accept the existence of the infinite, it was the *potential* infinite; and where the very idea of the infinite still seemed incoherent, it was the *actual* infinite.

It seemed necessary to accept the existence of the infinite given the infinitude of the natural numbers (which, for Aristotle, were abstractions from things and processes in the natural world[10]). But Aristotle felt able to accommodate this. Such infinitude did not mean that there were actually infinitely many numbers. Nor, certainly, did it mean that any individual number was infinite. (There was, for Aristotle, no such thing as the number of numbers.) It was to be understood rather in terms of there being no end to the process of counting.

Again, Aristotle felt bound to accept that space and time were infinite by division. Indeed he had a very elegant (though I do not say conclusive) proof that they were. It ran as follows.

Aristotle's proof of the infinitude by division of space and time: Consider two moving objects A and B, A being the faster. In the time t_1 that A moves a given distance d_1, B must move a shorter distance d_2. And in the shorter time t_2 that A moves the distance d_2, B must move a still shorter distance d_3. And so on *ad infinitum*.[11]

He was also convinced that matter was infinite by division. (We shall see shortly how he reconciled this with his earlier apparent rejection of the view.) But the infinitude by division of both space and matter he felt could be accommodated. For it could be understood in terms of there being no end to the process of successively dividing them. The infinitude by division of time posed more of a problem. How could a period of time be divided once it had elapsed? But here too Aristotle had an answer. He ingeniously reduced questions of temporal divisibility to questions of spatial divisibility, by invoking motion. (He believed that there was no time without motion.) We register the mid point of a period of time, for example, when we register the mid point of a line traversed, during that period, by an object moving at a constant velocity.[12]

These views about divisibility, involving, as they did, emphasis on the idea of a non-terminating process of division, revealed how closely related the infinite by addition and the infinite by division were for Aristotle. Indeed, given that he believed the world to be a finitely big place, he thought that the only available empirical example of the potentially infinite, aside from time itself, lay, precisely, in a never-ending process of division. Moreover, if such a process revealed an object to have parts (say) half its size, a quarter of its size, an eighth of its size, and so on – there always being more parts – this would be *like* revealing it to be infinite by addition. To this extent Aristotle was prepared not to distinguish between the two kinds of infinity.

In what ways, then, did Aristotle still feel the idea of the infinite to be incoherent? He remained convinced that there was something incoherent in the idea of a body's actually being divided into infinitely many parts.

But did he not hold body to be infinite by division?

He did, but he saw no tension here. There *seems* to be tension because of an ambiguity in the phrase 'infinitely divisible'. In one sense, Arstotle was denying that body was infinitely divisible. In another sense, he was affirming it. But his actual/potential distinction enabled him to distinguish between the two senses. What he was denying was that it was possible, by a process of division, to separate any body into infinitely many parts (for then it would have an actual infinity of parts). What he was affirming was that there could be no end to the process of dividing the body (which meant that it had a potential infinity of parts). For body to be infinite by division was for it to be infinitely divisible in the second sense.[13]

There is an elegant way of highlighting this ambiguity in terms of alternative structural parsings. Let us construe 'is divisible', for heuristic purposes, as 'is, in some possible situation, divided.' (For Aristotle, as has already been remarked, this was in turn akin to 'will, at some time, be divided.') And let us exploit the infinitude of the natural numbers. Then 'This body is infinitely divisible' can be glossed as either

(1) For every natural number n, there is a possible situation s, such that this body is divided into more than n parts in s

or

(2) There is a possible situation s, such that for every natural number n, this body is divided into more than n parts in s.

(1) means that however great a natural number you consider, a trillion say, this body can be divided into even more parts. (2) means that this body can be divided into a number of parts that is greater than any natural number you consider – and so infinite. Aristotle would have accepted (1), and he would have found (2) unintelligible.

It was thus that he solved the paradoxes of the infinitely small. What they show is always the incoherence of infinite divisibility in the objectionable sense; in its acceptable sense, it always remains intact. If Achilles runs straight from A to B, or overtakes the tortoise, say, which he can certainly do, then he does not actually perform infinitely many tasks or pass over infinitely many points. It is just that, however many divisions we recognize in his journey, we *can* always recognize more. (Insofar as it is *true* that Achilles' accomplishment is infinite, it is true, in the same sense, that the time he takes to carry it out is infinite – that is, potentially infinite by division. But Achilles cannot perform something like a staccato run, which involves an actual infinity of separate tasks.) Again, the time it takes for an arrow to fly through the air is not actually composed of infinitely many indivisible instances. It is just that there is no end to the instances we can recognize within it. (But the movement of the arrow must be understood relative to the whole time and not to the instances. It makes no sense, except perhaps derivatively, to speak of an arrow either as moving or as being at rest at any given instant.)

We now see, then, a beautifully integrated and coherent approach to the infinite in Aristotle. The problem that he faced had been basically overcome. For example, reconsider the three outstanding considerations in favour of the infinite that had seemed to pose such a threat to his position: (i), (ii), and (v). We can now see, as Aristotle himself went on to point out, that he could concede all three with equanimity provided that they were appropriately understood – or at least, we can see this subject to one proviso. There still seemed to be a problem about the commitment of geometricians to the infinitude of space by addition. Aristotle was not unduly perturbed by this. He pointed out, rightly, that what was assumed in some branch of mathematics about space and what was actually true of space need not be the same. He also made the point, noted in Chapter 1, §5, that there was no explicit reference to infinite space in geometry, only to points and finite lines. The fact remained that Aristotle was faced with a cleavage between his own view of space and the view embodied in contemporary geometry that meant that he could not regard geometry as supplying a straightforwardly true account of reality (still less an account that was true *a priori*). Still, his theory looked, in essence, very attractive.

He provided a pithy summary of it, quoted at the beginning of this chapter.[14] He likewise said:

> Something is infinite if, taking it quantity by quantity, we can always take something outside.[15]

All the elements are in these two quotations: the idea of temporal spread; the metaphor of reception; the finitude of each component within the infinite; and the importance of not repeating. The last deserves special mention. Aristotle wanted to emphasize that the untraversability of, say, a bezelless ring (or a uniform circular racecourse, to use our own earlier example), though it provided scope for an endless journey of sorts, was not his concern. Being able to go over the same ground again and again was no indication of anything genuinely infinite. The problem, of course, was that the ring (or the racecourse) was there 'all at once'.

A similar and crucial repercussion of his theory, which Aristotle was at pains to highlight, was that the *metaphysical* conception of the infinite (the complete, the whole, the unified) was precisely not his. Those who saw the infinite in these terms had it, he believed, completely back to front. 'The infinite turns out to be,' he said, 'the contrary of what it is said to be. It is not what has no part outside it that is infinite, but what always has some part outside it.'[16] This sounds cryptic. But essentially what he was urging was that the metaphysical conception of the infinite, such as he found in certain of the Eleatics and above all in Melissus, was a complete perversion of the true conception, which he now believed himself to have developed.

I said at the beginning of §2 that Aristotle appeared to abhor the mathematically infinite. We can now see how profoundly false such an appearance was. What he abhorred was the metaphysically infinite, and (relatedly) the actual infinite – a kind of incoherent compromise between the metaphysical and the mathematical, whereby endlessness was supposed to be wholly and completely present all at once. It was the mathematically infinite that he was urging us to take seriously. Properly understood, the mathematically infinite and the potentially infinite were, for Aristotle, one and the same. Far from abhorring the mathematically infinite, he was the first philosopher who seriously championed it. In so doing he recoiled from earlier thinking in such a way that he set the scene for nearly all subsequent discussion of this topic.[17]

5 A remaining difficulty

We close this chapter, however, with a difficulty that Aristotle's theory presented. He identified the infinite with the untraversable, and hence with that which cannot, at any point in time, have been traversed. This seems all very well when our attention is focused on the future. But what about the past? Aristotle believed that the past was infinite by addition. How could he reconcile this with its now having run its course – with its, apparently, having been traversed?

There is a related and curious asymmetry in our intuitions. Something infinitely old makes the mind boggle in a way in which something with an infinite future does not. Wittgenstein in a lecture once asked his audience to imagine coming across a man who is saying, '. . . 5, 1, 4, 1, 3 – finished!', and, when asked what he has been doing, replies that he has just finished reciting the complete decimal expansion of π backwards – something that he has been doing at a steady rate for all of past eternity.[18] There is a special way in which this story strikes us as absurd, a way in which the corresponding story about a man beginning to recite the complete decimal expansion of π and carrying on for ever does not strike us as absurd.[19] (It is a very good question what Aristotle might have said about this.)

Aristotle was not entirely unaware of this difficulty. He insisted, for example, that the succession of time did not involve a steady accumulation of persisting things:[20] this meant that past time did not, in any *obvious* way, present us with an actual infinity. But he believed that time had never begun, and that there had always been motion (the revolution of the heavens).[21] And it is hard not to see already in this – the past now *being* past – a presentation of the actual infinite. There may be sophisticated views about time that would have enabled Aristotle to circumvent this difficulty. He may even have held such views.[22] But the difficulty, as we shall see, was to prove a significant factor in later assimilation of his ideas.

CHAPTER 3

Medieval and Renaissance Thought[1]

A finite intellect ... cannot by means of comparison reach the absolute truth of things. Being by nature indivisible, truth excludes 'more' or 'less', so that nothing but truth itself can be the exact measure of truth. ... In consequence, our intellect, which is not the truth, never grasps the truth with such precision that it could not be comprehended with infinitely greater precision. (Nicholas of Cusa)

1 The Greek legacy: reactions and developments

Two things dominated medieval and renaissance philosophy in general, and philosophical thought about the infinite in particular: the legacy of the Greeks; and religion. Religion in this context virtually meant Christianity, but not exclusively so. And the legacy of the Greeks was eventually to become, more than anything else, the legacy of Aristotle – albeit tempered and informed by strands of Platonism.

Before that, however, there was something of a reaction against Aristotle and a reversion to the ideas of Plato – albeit tempered and informed by strands of Aristotelianism. Plotinus (205–270) played a major role here. He was probably born in Upper Egypt and may have been a Hellenized Egyptian. He was not a medieval thinker. He stood, rather, at the end of antiquity. But this is an apt point at which to consider him.

He drew heavily on the ideas of Plato, founding what became known as Neoplatonism. This was to have a profound influence on Christian thought. First and foremost he wanted to resurrect (something like) the very radical appearance/reality distinction that had been integral to Plato, and before him Parmenides and the Eleatics. He believed in an utterly transcendent realm of being that underlies and sustains, yet is quite separate from, all that we directly experience. It was hostility to this very belief that had been at the roots of Aristotle's naturalistic espousal of the mathematically infinite, and his repudiation of the metaphysically infinite. By readopting the belief, Plotinus was in a position to upturn much of what Aristotle had argued for and to rehabilitate the metaphysically infinite. He

45

referred to the underlying reality sometimes as The One (in an Eleatic vein), sometimes as The Good (in a more Platonic vein), and sometimes as God, but also as infinite, and he explained this in such a way as to make clear that he meant, very definitely, metaphysically infinite. He called it self-sufficient, perfect, and omnipotent, a complete and pure unity, utterly beyond our finite experience. He also said that it was 'supremely adequate, autonomous, all-transcending, most utterly without need.'[2] Sometimes he spoke of it in a Parmenidean way, implying that it had internal limits. 'Its manner of being is settled for it,' he said, 'by itself alone.'[3] But elsewhere he emphasized its lack of limits, either external or internal.[4] Indeed, in line with this, he insisted that all our attempts to talk about it or define it were strictly speaking, and inevitably, inadequate. Thus, in truth, it even transcended such descriptions of it as 'The Good' or 'God'. Its ineffability meant that we had to be content with mystical insight into it. He nevertheless tried to convey as much as possible in words. And in so doing he supplied one of the first explicit identifications of the infinite with God. In this his thinking marked something of a turning point. No longer in the history of philosophy would there be, as there had been among the Greeks, a tendency to hear 'infinite' as a derogatory term. Henceforth, quite the opposite.

Significantly, however, Plotinus combined all of the above, which committed him so fundamentally to what Aristotle had so fundamentally opposed, with a thoroughly Aristotelian conception of infinitude in the *sensible* world. (This – almost impudently – eclectic strategy was later to be emulated by many others with similar metaphysical views, and we shall see the same pattern several more times in this chapter.) Thus he denied that there was infinite number, and he denied that there was any actual endless extension among sensible beings (although matter, in itself, lacked a *peras* in the sense of being indeterminate). But he recognized the potential or mathematical infinitude of time, urging that, over against the metaphysically infinite (or in his words, 'the immediately Infinite'), there must be 'that which tends ... to infinity but by tending to a perpetual futurity.'[5]

Something very like this was to be found in St Augustine (354–430), who was born in Thagaste, in northern Africa, and became the bishop of Hippo. Augustine was one of the first and greatest of medieval thinkers. He was deeply influenced by Neoplatonism, much of which he attempted to integrate with Christianity. (He did a great deal to propagate the general impact of the former on the latter.) He too resisted the heavy emphasis on experience that had motivated Aristotle, and this enabled him to acknowledge, with Plotinus, an infinity that was more than just potential. The point was (as Aristotle had indeed helped to show) we could have no direct experience of it. On Augustine's conception, God was both actually infinite and transcendent. He was infinite in a way that enabled Him to know the totality of natural numbers for example (though interestingly,

Augustine believed that, in knowing this totality, God 'made it finite': it was bounded by His knowledge, in a sense that we cannot express).[6] Like Plotinus, however, Augustine took up the Aristotelian theme that, although time was infinite, its infinitude was less than this: it was never present 'all at once'. Indeed he reinforced this idea by pressing the view of time it presupposed, namely that time itself was never present all at once. The future did not really exist – was not already given – in the present. (Otherwise the infinitude of time would already be given and the potential infinite would be a kind of actual infinite.) God's 'eternity', on this conception, insofar as it *was* present all at once, had to be something transcending mere existence at all times.[7] This was a view that was embraced by many later Neoplatonist theologians, Boethius for example.

Both Plotinus and Augustine, then, preserved certain Aristotelian insights about the infinite despite a fundamental opposition. One of the characteristic features of this period, even before Aristotle became any kind of authority, was the extent to which attempts were made to integrate his views with cherished beliefs. This was particularly true of religious beliefs, and it was true more particularly still, though not exclusively, of the main tenets of Christianity. When it came to integrating Aristotle's views with these, two difficulties arose that were to have a major bearing on discussion of the infinite throughout the period.

(i) Aristotle had denied the possibility of an actual infinity in the world of nature. But this had to be reconciled with belief in God's omnipotence. Could not God create something actually infinite if He wanted to?

(ii) Aristotle's theory had already presented a difficulty about the past, as we saw at the end of the last chapter. Aristotle had gone some way towards mitigating this by denying that the succession of time involved a steady accumulation of persisting things. But belief in the immortality of the soul undercut this move, and thus exacerbated the original difficulty. Immortality meant that an infinitely old world would, or at least could, produce an actual infinity of souls.

Some of the early Neoplatonists denied that it *would* do so, by appeal to reincarnation.

Later thinkers who accepted the literal truth of Jewish, Christian, or Islamic orthodoxy (in particular, concerning creation) circumvented (ii) by simply surrendering the Aristotelian belief in an infinitely old world. This was Augustine's move.[8] It might be thought that this still left a problem about infinite past time. But Augustine, like Aristotle and many others of this period, believed that time involved activity: if the world was finitely old, then likewise time itself.[9]

John Philoponus (late fifth century to after 550), a Christian who was probably born in Caesarea and held a chair in Alexandria, also believed the world to be finitely old. But he was a stern critic of Aristotle and felt less conflict than most. For him, (ii) was simply (the basis of) one of a whole battery of arguments that reinforced Aristotle's own original quandary and told against an infinitely old world. Some of the rest of these arguments are significant not only because of how often they were later rehearsed but also, in the context of this enquiry, because they constitute the first clear allusion to the paradoxes of the infinitely big.[10] For example, one of these arguments can be recast as follows.

> *Philoponus' argument*: However many men there were before Socrates, there have been more by now; again, however many months the world has existed, it has existed over thirty times as many days. So if the numbers here were infinite, one infinity would be greater than another. But this is absurd. So the numbers must be finite. The world must be finitely old.

Two Islamic Persian philosophers, Avicenna (980–1037) and Abū Hāmid Muhammad Ghazālī (c. 1059–1111), also took up discussion of (ii). Avicenna tried to circumvent it by arguing that an actual infinity of souls was unproblematical provided they were not ordered in any way. (The paradoxes of the infinitely big all seem to trade on a natural ordering of the sets being considered.[11]) But Ghazālī replied that the souls *would* be ordered – temporally, according to when they were created. His own solution, given that he was also aware of the kinds of problems with an infinitely old world that Philoponus had set forth, was the same as Augustine's and Philoponus': to accept religious orthodoxy, and simply deny that the world was infinitely old.[12]

2 Aquinas

Eventually in the Middle Ages, as I have already said, Aristotle became something of a philosophical authority. This was due, above all, to the great Italian theologian St Thomas Aquinas (c. 1224–1274). Aquinas was perhaps third only to Jesus Christ and St Paul in the extent to which he shaped subsequent Catholicism. The main philosophical strand in his thinking was Aristotelianism.

Nevertheless, because of his commitment to Christianity, he parted company with Aristotle on the same fundamental issue as the Neoplatonists. He believed in a metaphysical infinitude, the metaphysical infinitude of God, Whom he held to be self-subsistent and perfect. He did not believe that God was *mathematically* infinite, since this would have meant His having parts and therefore being imperfect.[13] (Compare this with Melissus (see above, Chapter 1, §3).) Belief in God's metaphysical infinitude, and in particular in His perfection, did, of course, pose a familiar theological problem: how was evil possible? Aquinas' reply was that God allowed evil in order to bring good out of it.[14]

These beliefs apart, his views about infinity were very Aristotelian. He believed that nothing in creation was metaphysically infinite, by definition; to be created is, precisely, not to be self-sufficient. He also believed that nowhere in creation was there an actual mathematical infinite, either as a magnitude (a property of a single thing) or as a multitude (a property of a plurality of things). But he accepted the potential infinite, in essentially Aristotelian terms and relying on essentially Aristotelian arguments. There was one very interesting new twist: he thought that, because we could know what (say) greenness was, and could therefore recognize indefinitely many things as green, we had a kind of infinite power; but he did not think that this threatened his views about the infinite in creation (any more than the fact that we could count indefinitely).[15]

How, then, given that he was so thoroughly immersed in both Christianity and Aristotelianism, did he cope with the two difficulties outlined above?

He coped with (i) simply by pointing out that not being able to do the impossible was no limit on God's omnipotence. (One would similarly have to deny that an omnipotent being could create something uncreated.)

His reaction to (ii) was fascinating. Aquinas was in the same position as Augustine, Philoponus, Ghazālī, and others: he wanted to reject Aristotle's belief in an infinitely old world on quite independent grounds, accepting the scriptural account of its creation. This meant that the one thing that had seemed to be the Achilles' heel of Aristotle's theory of infinity (in nature) was the one thing that Aquinas wanted to repudiate anyway. Yet ironically, he was at great pains to defend Aristotle on this very point (not, indeed, by arguing that the world *was* infinitely old, but by trying to show that there was nothing incoherent in Aristotle's supposing so). This was because, although Aquinas did believe in the scriptural account of creation, he was convinced that the truth of the account was revealed. That is, he was convinced that we needed to appeal to the Scriptures in order to find such truth; it was not something we could discover for ourselves by reason. Consequently he made various attempts to circumvent Aristotle's difficulty, both in its original form and as exacerbated by belief in the immortality of the soul. He pointed out that even if the world had existed for infinitely many days, no one of these days would be infinitely far away (which would, he admitted, be incoherent). And he rebutted the threat of immortality by appeal to the possibility of reincarnation, which the early Neoplatonists had latched on to. There were various other moves he made. The net effect was that his arguments concerning the infinite looked even more Aristotelian than the thoroughly Aristotelian position he ultimately sought to defend.[16]

The Infinite

3 Later developments: the mathematically infinite[17]

Consensus. The categorematic/syncategorematic distinction

The thirteenth and fourteenth centuries saw interesting developments and growing sophistication concerning the mathematically infinite. It was generally accepted, on Aristotelian grounds, that it was wrong to think of an infinitely divisible quantity (a line or solid, say) as actually being composed of infinitely many infinitesimal parts. The Scottish philosopher and theologian John Duns Scotus (c. 1266–1308) used Figure 3.1, which shows two concentric circles, to urge this point. He argued as follows.

Duns Scotus' argument: Suppose that lines were composed of infinitely many infinitesimal points. Then the the two circumferences in Figure 3.1 would be composed of the same number of points, for, as we can see from the figure, the points can be paired off. This in turn means that the two circumferences would be equal, which they manifestly are not. We must therefore reject the original supposition. Lines are not, after all, composed of infinitely many infinitesimal points.

Comment: If we think that lines *are* composed of infinitely many infinitesimal points, then we can regard this as yet another paradox, sharing features with the paradoxes of the infinitely small, and of the infinitely big.

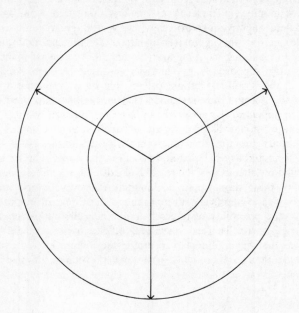

Figure 3.1

50

Despite this argument, Duns Scotus was happy to accept that infinite-simals such as points and surfaces could genuinely exist in their own right (for example, a surface could have a colour), whereas William of Ockham (c. 1285–1349), the influential English philosopher, held that they were 'pure negations'. On William's view, which was at once very Aristotelian and very modern in its sophistication, talk about infinitesimals was just a *façon de parler*, abbreviating talk about ordinary finite solids: to say that a sphere touched a plane at a single point, for example, meant that (a) the sphere touched the plane, and (b) there was no limit to how small a part of the sphere you could consider such that that part, and none of the rest of the sphere, touched the plane. (This is reminiscent of the sophistication of Eudoxus and Archimedes (see above, Chapter 1, §5).)

This kind of emphasis on careful speech and the precise reformulation of loose idioms, aimed at combating confusion, was a hallmark of the period. Thus it was that, during this period, an important new distinction was drawn. Recall how, in §4 of the last chapter, the ambiguity in the sentence, 'This body is infinitely divisible,' was highlighted in terms of alternative structural parsings. Something similar happens whenever 'infinite' or its cognates (and likewise 'indefinite') is used to qualify a capacity in this way. If I say, 'This body can move infinitely fast,' (or 'indefinitely fast,') I may mean that it is capable of achieving an infinite speed, or I may mean that there is no limit to the finite speeds which it is capable of achieving. The new distinction provided a way of registering the different interpretations in these and related ambiguities. Peter of Spain (c. 1220–1277), who was born in Lisbon and became Pope John XXI, was the person who paved the way. Having signalled one such ambiguity, he said that, when 'infinite' was being used in the first way, it was being used *categorematically*, whereas when it was being used in the second way, it was being used *syncategore-matically*, and he went on to try to explain the distinction. His explanations were later improved and refined by others, notably the Frenchman Jean Buridan (c. 1295–1356) and the Italian and Augustinian Gregory of Rimini (c. 1300–1358). The gist of their explanations was encapsulated in those two parsings in Chapter 2. Roughly: to use 'infinite' categorematically is to say that there is something which has a property that surpasses any finite measure; to use it syncategorematically is to say that, given any finite measure, there is something which has a property that surpasses it.

Later, this distinction was exploited more and more. Appeal to it became commonplace. It not only facilitated disambiguation; it was also useful in classifying unequivocal uses of 'infinite'. Furthermore, it clearly bore on Aristotle's actual/potential distinction. It would be an oversimplification to say that it was just intended to *be* it, in a different guise, or that it was intended to usurp its role. (Peter had explicitly denied any coincidence.) But it did do very similar work in an interestingly new way. More than that, given sufficiently barbaric regimentation, the actual/potential distinction could be subsumed under it. For – and this is again rough – to use 'infinite' in order to refer to an actual infinite is to say that there is some *time* by which a given magnitude surpasses any finite measure (the magnitude might be the number of divisions in a body, for example); and to use 'infinite' in order to refer to a potential infinite is to say that, given any finite measure, there is some time by which a given magnitude surpasses it. By way of illustration, consider the following application of the new distinction in a temporal context, noted by Gregory. If I say, 'An infinity of men will be dead,' and use 'infinity' categorematically, then I mean that there will come a time when infinitely many men are dead; there will then be an actual infinity of dead men. If I say the same thing, and use 'infinity' syncategorematically, then I mean that there is no end to the number of men who will, each in his own time, be dead; there is a potential infinity of dead men. This explains, I think, why so many philosophers have thought that there was something deeper and more abstract underlying straight-forward temporal accounts of the actual/potential distinction. It seems they were right. There *is* something – something grammatical. (Working with the new distinction also has the advantage that one can avoid the false implication in Aristotle's terminology, noted by Aristotle himself, that what is potentially infinite must be capable of being actually infinite.)

Controversy

What we have just seen was a broad measure of consensus, of a roughly Aristotelian kind. Controversy broke out concerning the actual infinite by addition. This was sparked off by the two difficulties (i) and (ii), which I outlined in §1.

Many held on to the Aristotelian conviction that the actual mathematical infinite was incoherent. For example, Richard of Middleton (died c. 1300), who is believed to have been either French or English, accepted that it was a contradiction in terms, almost like the complete incomplete. (This view was later eloquently defended by Peter Aureol.) Richard's solution to (ii) was the by now familiar one of simply denying that the world was infinitely old (or even that it could have been).

The Englishman Walter Burley (1275 to after 1343) likewise denied the coherence of the actual mathematical infinite. His main difficulty was (i). For he believed that God could create something indefinitely big. He tried to alleviate the tension by disambiguating this claim: it was not that God could create something whose size was actually greater than any finite limit (which would require the existence of an actual mathematical infinite); it was just that there was no finite limit to the size of that which God could create. Burley's solution was obviously an application of the categorematic/syncategorematic distinction.

At the same time, however, an opposed and very new trend was developing, namely to undercut (i) and (ii) much more directly, by shrugging off Aristotelian qualms about the actual mathematical infinite. To do this was, by now, to flout received and time-honoured wisdom. But we must not exaggerate the extent to which Aristotle was able to have things his own way. Remember how many of his own objections to the actual mathematical infinite had been empirical. Many of these would now have seemed outdated. Moreover, the more conceptual of his objections had depended on a particular temporal conception of the infinite that was beginning to lose its appeal. In fact, now that the paradoxes of the infinitely big had become familiar, the most serious objection to the actual mathematical infinite, at least by addition, seemed to lie in them. But it was felt by many that they could be dealt with.

It was Duns Scotus who prepared the way for this new trend. (His discussion of the concentric circles meant that he had already dealt with one such paradox.) First he argued that our inability to conceive an actual mathematical infinity merely reflected our own finite limitations. Then he tried to rebut the paradoxes of the infinitely big by denying that comparisons of size so much as made sense between infinite quantities; paradoxes arose because we were trying to understand the infinite in finite terms.

But it was Gregory who really developed the trend. His attitude was somewhat different from that of Duns Scotus, though. On his view it made *sense* to say that one infinite quantity was smaller than another, but we had to be careful to specify just what sense. We were confronted with paradoxes if we did not do so. Various distinctions that did not need to be drawn in the finite case did in the infinite case (which was after all only to be expected). For example, the paradoxes of the infinitely big showed that in the infinite case it was possible for one collection to be contained within another *and* for their members to be susceptible of a pairing off. In such circumstances, the former collection would be smaller than the latter in one sense, but not in another (a distinction that could not arise in the finite case). Again, the former collection would have fewer members in one sense, not in another. But there was no threat of contradiction here. It just meant that we had to observe the relevant distinctions and always make clear exactly what we meant. (William of Ockham took a very similar line. It had also been anticipated by Henry of Harclay.)

Not only did Gregory believe that we could coherently discuss the actual infinite in this way, he used the materials that had made up earlier quandaries to argue for its possibility.

Gregory's argument: If God can endlessly add a cubic foot to a stone – which He can – then He can create an infinitely big stone. For He need only add one cubic foot at some time, another half an hour later, another a quarter of an hour later than that, and so on *ad infinitum*. He would then have before Him an infinite stone at the end of the hour.

Gregory also claimed that, in the sense that they could be appropriately paired off, all infinite quantities were the same size. He lacked a proof of this. He may have thought it obvious. It is not. We shall return to this issue later (see below, Chapter 8, §3). Finally, he even challenged the orthodoxy on infinitely divisible quantities. He believed that these *were* actually composed of infinitely many infinitesimal parts.

Controversy thus reigned. Jean Buridan (among others) continued to find the actual mathematical infinite highly suspicious. For example, he pointed out the problems that arise if we entertain the possibility of a spiral extending (actually) infinitely far inwards. Turning to Gregory's argument, he tried to rebut the conclusion by invoking the categorematic/ syncategorematic distinction. He argued that God could, acting ever more quickly, create a stone of arbitrary finite size within an hour, but that He could not complete the infinite task specified.

Much later, the great Italian astronomer and physicist Galileo Galilei (1564–1642) was to join in the controversy. He formulated his own paradox. This was just like the paradox of the even numbers except that it focused instead on the squares.

Galileo's paradox: Figure 3.2 shows that we can pair off the natural numbers with those that are squares. This suggests that there are just as many of the latter as of the former. But since not all the natural numbers are squares, it seems obvious that there are fewer of the latter.

Comment: What makes Galileo's paradox particularly striking, as he went on to point out, is the fact that, as larger and larger initial segments of the sequence of natural numbers are considered, the proportion which are squares gets closer and closer to 0. For example, half of the first four natural numbers are squares, but only a tenth of the first hundred and only a thousandth of the first million.

In discussing this paradox Galileo reverted to the line that Duns Scotus had taken. He argued that we could not help thinking about the infinite, but that it transcended our finite understanding; and the reason we became ensnared in contradiction was that we tried to apply to it concepts which should only be applied to the finite. (This is not in fact so different from Gregory's view. He too would have said that there were certain concepts, in which various distinctions were obliterated, that were perfectly service-

Figure 3.2

able in the finite case but that ceased to be of any use in the infinite case.)

Galileo also discussed Duns Scotus' concentric circles. He insisted that they *were* composed of infinitely many infinitesimal points, which could be paired off, but that the longer line contained infinitely many infinitely small gaps not in the shorter line. This too, however, was something we could not hope to understand with our finite intellectual resources. One could talk about the potential infinite instead of the actual infinite as a way of easing the mystery: but he did not think that this addressed the real problems, nor that it was necessary.[18]

4 Nicholas of Cusa. The end of the Renaissance

In the fifteenth century Nicholas of Cusa (1401–1464), who was born at Kues and spent much of his life in Germany, provided ample evidence that metaphysical concerns were still alive and that the prevailing Aristotelianism had not in any sense gained a stranglehold. Nicholas returned to Platonic, nay Eleatic, views of the most radical kind, revitalizing a powerful metaphysical conception of the infinite. The upshot was something highly reminiscent of Plotinus (bringing this chapter, in a way, full circle).

Nicholas described the infinite sometimes as God, sometimes as the truth, sometimes as the Absolute Maximum, while at the same time insisting that all such descriptions contained an element of falsification. What we had to realize was that we could never grasp it; nor could we compare it or relate it to anything that we could grasp. True wisdom consisted in recognizing our own essential ignorance.

What we had, on Nicholas' view, was a point of view that was, in the words of Rawson's hymn, 'crude, partial, and confined.' We could continually improve this point of view and approximate more nearly to a grasp of the truth. But the point of view could never fully attain to the truth without actually becoming it. And becoming the truth did not admit of degrees, any more than the truth itself did. Here Nicholas drew an analogy with Eudoxus' and Archimedes' method of exhaustion (see above, Chapter 1, §5). Our point of view was like a polygon, and the truth a circle in which it was inscribed: increasing the number of sides of the polygon reduced error but never produced the circle. The finite could never be made into the infinite.

One of the limitations on our point of view was that we could not tolerate contradiction. But the truth, or the metaphysically infinite, was all it could possibly be, and there was no reason to suppose that it was free of contradiction. It was beyond all finite categories. Thus it was as much the Absolute Minimum as the Absolute Maximum. At the level of the infinite, finite distinctions collapsed. Nicholas had a variety of mathematical examples by means of which he illustrated this. The greater a circle, the less the curvature of its circumference. At the limit, Nicholas argued, an infinite circle is an infinite straight line. It inevitably followed that the infinite was for us, given our finite modes of thinking, indescribable and indefinable. We could perhaps *name* it, but insofar as our name purported also to describe it (as 'the Absolute Maximum' did), then it contained – to repeat – an element of falsification.

Of course, this meant that Nicholas faced the fundamental paradox of thought about the infinite. With what right could he say any of this? What reason did he have even for believing in the existence of the infinite?

Nicholas was convinced that, through faith, we could come to a direct, intuitive, ineffable awareness of the infinite. His various attempts to put this into words seemed to be grounded in the conviction that our own finitude only made sense in terms of, and in contrast with, the infinite. For Nicholas believed that finite beings were partial appearances of the infinite, and that God was as immanent as He was transcendent. Nicholas' view of the natural world reflected this. Insofar as it was just a partial appearance of the infinite, it was not itself infinite. It was, so to speak, finite from God's point of view. From our point of view, on the other hand, it was without external limits and incapable of extension, or, in Nicholas' phrase, 'privatively infinite'. This was Nicholas' way of introducing the mathematically infinite into the realm of appearances, which constituted a kind of image, spread out in space and time, of the underlying metaphysical infinitude.[19]

The Italian Giordano Bruno (1548–1600) was greatly influenced by Nicholas and had a very similar vision. He believed in a transcendent, self-sufficient, ineffable Deity Who contained contradictions within Himself, and Who expressed Himself in a spatially, temporally, materially infinite universe. Bruno believed that we could, by appropriately extending ourselves, gain a kind of gnostic union with this Deity – the truly infinite. He also argued for the mathematical infinitude of space on independent grounds, in much the same way that Archytas had done (see above, Chapter 1, §5) – an argument that had also subsequently been taken up by Lucretius incidentally. And he argued that this entailed the mathematical infinitude of what was *in* space, in much the way that Aristotle had done (see above, Chapter 2, §2).[20]

Bruno's death – he was burnt at the stake by the Inquisition – is sometimes regarded as a convenient point by which to mark the end of the Renaissance. It was also nearly two thousand years after the time of Aristotle. Two things have perhaps emerged in this lightning sketch of that period: the enormous influence that Aristotle exercised; and, despite that and largely because of the impact of religion, the continuing vitality of issues that he might have hoped to have exorcized.

CHAPTER 4

The Calculus[1]

And what are ... fluxions? The velocities of evanescent increments? And what are these same evanescent increments? They are neither finite quantities, nor quantities infinitely small, nor yet nothing. May we not call them the ghosts of departed quantities?　　　(George Berkeley)

1 The fundamental principles of the calculus

There is a cluster of mathematical problems which at first sight seem quite disparate but which come together in a remarkable way. They have a history going back at least as far as Eudoxus and Archimedes, with their method of exhaustion, and continuing to the present day. This history is a vital component strand within the broader history of the infinite. I turn to it now, because some of the most significant breakthroughs that go to make it up were made in the seventeenth century.

What are the problems to which I refer? They fall into two groups. Those in the first group all concern curves. They include: how to determine the area of a curved figure (a figure bounded by curves); how to determine the slope of the tangent to a curve at a point; and suchlike. The problems in the second group concern the idea of the continuous variation of one quantity with respect to another. (For example, if an object moves straight from A to B, accelerating all the time, then, during that period, both its distance from A and its speed increase continuously with respect to time.) These problems include: how to analyze such continuous variation; how to determine its rate; and suchlike.

All of these problems come together in that branch of mathematics known as *the calculus*. The first step towards a grasp of the calculus is to understand what is meant by a *graph*. Suppose two lines, or *axes*, at right angles to each other, represent the different possible values of two quantities. Then a point on the same plane as them can represent a pair of these values, as determined by its distance, in the relevant direction, from each axis (see Figure 4.1). The two values are known as the point's *co-ordinates*. A graph can be thought of as a set of such points. (For convenience, I am adopting a somewhat broader definition of a graph than is standard.)

The continuous variation of one quantity with respect to another can be represented by a graph. It will take the form of an unbroken line (unbroken, because of the continuity). For example, suppose that an object moves during a two-second period in such a way that after x seconds it has moved x^2 feet, $0 \leq x \leq 2$. (For instance, after $1\frac{1}{4}$ seconds it has moved $1\frac{9}{16}$ feet.) Then the object's movement is represented by the graph in Figure 4.2. A graph of this kind not only represents how one quantity varies continuously with respect to another, it is also a curve. Thus graphs are a vital point of contact between the two groups of problems with which we started.

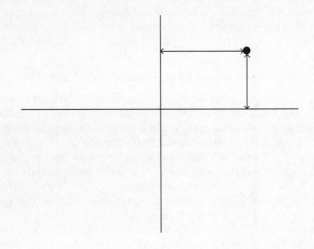

Figure 4.1

More generally, they are a vital point of contact between geometry and analysis. (Analysis is the theory that deals with real numbers, as defined in Chapter 1, §2.) A circle, for example, is a graph. Its centre can be taken as the intersection of two axes; and given any point on the circle, its co-ordinates x and y will satisfy the equation $x^2 + y^2 = r^2$, where r is the radius of the circle, and where x, y, and r are all real numbers. This is due to Pythagoras' theorem, and is illustrated in Figure 4.3. Application of these methods is known as *analytic geometry*. It is nothing less than the casting of one whole body of mathematics in terms of another (geometry in terms of analysis). In its beauty, depth, and power, it is one of the greatest monuments to mathematical excellence.

Some of the basic ideas of analytic geometry were familiar even in

The Calculus

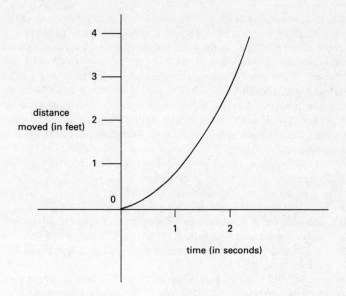

Figure 4.2

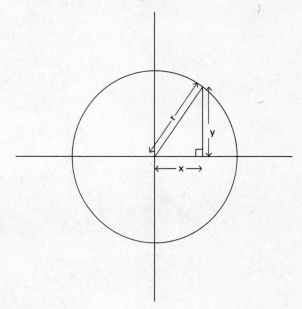

Figure 4.3

antiquity, and pioneering work in its direction had been carried out in the fourteenth century by Oresme. But its full development came in the seventeenth century. It is normally associated with the great French mathematician and philosopher René Descartes (1596–1650), and co-ordinates are often called Cartesian co-ordinates after him. But earlier unpublished work of a similar kind had been done by another Frenchman, Pierre de Fermat (1601–1665). Each of them produced work of genius.

We can exploit some of this work as we return to the question of how the two original groups of problems are related. The example of the moving object can act as a useful prop. The average speed of the object throughout the two-second period is two feet per second. It moves four feet in two seconds. But suppose we want to know its speed *at a particular time* (as it were, the reading on its speedometer at that time). In other words, suppose we want to know at what rate the distance it has moved is changing, with respect to time, *just then*. Geometrical intuition tells us that this is a matter of how steep the graph is at the relevant point; and that this is in turn a matter of the slope of the tangent to the curve at that point. But how do we determine this?

Consider a particular point P on the curve, representing the time after x seconds, when the object has moved x^2 feet. And consider the straight lines joining P to points further along the curve, as illustrated in Figure 4.4. The

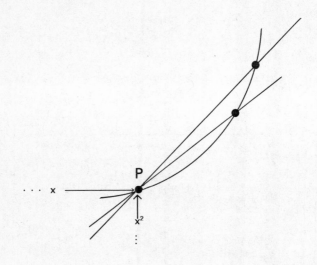

Figure 4.4

closer these points are to P, the more nearly the lines approximate to the tangent at P. This idea of successive approximations is reminiscent of Eudoxus' and Archimedes' method of exhaustion, discussed above in Chapter 1, §5. And it is tempting here, as it was then, to think of the limit in infinitary terms. It is tempting, that is, to say that the tangent at P is the line joining it to a point infinitesimally close to it. We have been given no license for succumbing to this temptation. But let us continue to think in these terms, for heuristic purposes, as we did before, to see what comes of it. We can then construct the following argument.

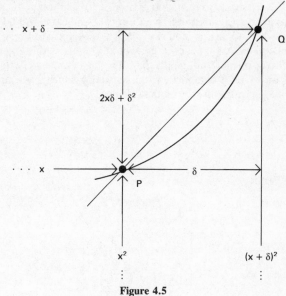

Figure 4.5

Let Q be a point further along the line from P, representing the time after $x + \delta$ seconds, when the object has moved $(x + \delta)^2$, or $x^2 + 2x\delta + \delta^2$, feet, as illustrated in Figure 4.5. The slope of the line joining P to Q is, to put it intuitively, the number of units it rises for each unit it goes along. But if we compare the co-ordinates of P with those of Q, we can see that it rises $2x\delta + \delta^2$ units when it goes along δ units. So the slope of the line is the result of dividing the first of these by the second, which is $2x + \delta$. Now if Q is infinitesimally close to P, δ can be discounted and this is in turn equal to $2x$. So the slope of the tangent at P is $2x$.

Comment: If we now apply this result to the original problem, then we can conclude that the speed of the object after, say, 1¼ seconds (to use the same example as before) is 2½ feet per second.

Now it is clear that something is wrong with this argument. When Q is infinitesimally close to P, δ is supposed to be sufficiently like 0 for $2x + \delta$ to be equal to $2x$ and yet sufficiently different from 0 for there to be no problems about dividing by δ. (Division by 0 is illegitimate. Why? Because if it were allowed, we could prove, say, that $2 = 3$, given that $2 \times 0 = 3 \times 0$.) It might be said that δ is greater than 0 but less than any positive number (just as it might be said that the angle between the curve and the tangent is greater than 0 but less than any positive number). However, pending some explicit account of what this is supposed to mean, it is quite senseless. There seems to be no alternative but to admit that the argument involves us in an illicit attempt to have our cake and eat it. We take δ to be equal to 0 when it suits us, and not when it does not. However, like our 'reconstruction' of Archimedes' argument in Chapter 1, this argument has some intuitive appeal. And, *unlike* that, it is not an unfair reflection of the reasoning on which it is modelled, the reasoning which was employed in the early stages of the history of the calculus.

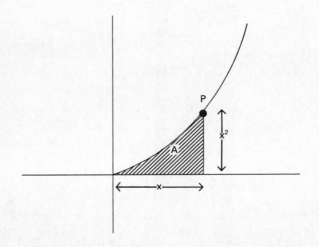

Figure 4.6

Before we look further into that history, let us glance at another application of the same kind of reasoning. Reconsider the point P. Suppose we want to know the area A of the figure bounded by the curve, the time-axis, and the perpendicular to the time-axis from P, as illustrated in Figure 4.6. (In accordance with the principles of analytic geometry, we shall think of areas as real numbers.) Then we can argue, much as we did before, as follows.

An infinitesimal increase in x will correspond to an infinitesimal increase in A taking the form of an infinitesimally thin strip whose height is x^2. Another way of putting this is as follows: the result of dividing an infinitesimal increase in A by the corresponding infinitesimal increase in x is x^2 – just as it was 'proved', in the argument above, that the result of dividing an infinitesimal increase in x^2 by the corresponding infinitesimal increase in x (which is the slope of the tangent at P) is $2x$. So to see how to express A in terms of x we must take the same step as we took when seeing how to express the slope of the tangent at P in terms of x, only backwards; backwards, because this time what we know is the *result* of the relevant division. It turns out – the details are beyond the scope of this book – that $A = \frac{1}{3}x^3$.

The connections here are remarkable. The fact that we have to take the same step in both arguments, only once backwards, is, very roughly, the import of what is known as the fundamental theorem of the calculus. It highlights a deep and unsuspected connection between two of our original problems. The 'forward' step is the characteristic step of what is called the *differential* calculus. The 'backward' step is the characteristic step of what is called the *integral* calculus.

But we must not get carried away. For all its depth and beauty, the reasoning here is, as we have seen, fundamentally flawed. It rests on a certain notion of an infinitesimal difference (as not quite nothing, but not quite something either), and this notion is ultimately incoherent. Let us turn now to the history of the calculus, to see how this problem animated it and was eventually solved.[2]

2 A brief history of the calculus

Leibniz and Newton

The calculus was invented, quite independently, by two men: the Englishman Sir Isaac Newton (1642–1727), arguably the greatest scientist of all time; and the brilliant German mathematician and philosopher G.W. Leibniz (1646–1716). This is not quite the coincidence that it sounds. Neither man was working in a vacuum. Work paving the way for them had been carried out by many earlier thinkers. In particular, Descartes and Fermat again deserve special mention.[3] Nevertheless, by generalizing and developing the methods sketched in the previous section, and discovering deep principles of connection between them, Newton and Leibniz can each be said to have established, in his own distinctive way, the calculus proper. Although neither of them produced a system that was completely original, nor, for that matter, completely immune to criticism (as we shall soon see), they are rightly acclaimed for the greatness of their achievement.

Newton arrived at his dicoveries in the mid 1660s, about ten years earlier than Leibniz, but he published later. In fact, some of his most important contributions to the field were published posthumously. There were, between their disciples, if not between them, fierce and vitriolic debates about who really got there first and who deserved the greater credit. It does not matter. More interesting were the differences between their approaches.

There were some trifling differences in terminology. We owe to Leibniz the actual expressions 'differential calculus' and 'integral calculus'. Newton described his system as the method of fluxions. He defined a *fluent* as a quantity that varies over time, and a *fluxion* as the rate at which it does so. Leibniz became interested in these issues because of his conviction that all change in nature is continuous, Newton because of how it all bore on his scientific discoveries, particularly in mechanics. Leibniz had the more analytic emphasis, Newton the more geometrical. Leibniz' system was the more elegant and the more versatile. It used notation which was slicker and easier to handle, and which is still used today. (This was no doubt a reflection of his life-long philosophical ambition to discover a '*Characteristica Universalis*', a clear and precise symbolism for the expression of any possible thought that would enable all questions to be settled by calculation.) Leibniz' system was also the more successful and influential, at least outside Britain.

One thing that they had in common, however, was that they both made use of the infinitesimally small. The sketch of their methods in the previous section does not, in either case, count as a gross caricature. Leibniz was well aware of the problems that this raised. His reaction was to urge us not to take talk of infinitesimal quantities literally. We could think of it as just a *façon de parler*, or as a 'useful fiction', to be justified by appeal to its enormous utility. (I shall return to the question of its utility below.) Still, if such talk made *sense* – Leibniz may have been saying that it did not – this left open the question of what sense. Someone might say, for example, that talk of the average parent was just a *façon de parler* (which is why the average parent, unlike any ordinary parent, can have 2.4 children); but they would typically back this up with some account of how it was to be paraphrased in terms of talk of ordinary parents, thereby giving its sense. William of Ockham, as we saw in the last chapter, §3, said that talk of points was just a *façon de parler*; but he backed this up with an account of how it was to be paraphrased in terms of talk of finite solids. Was talk of infinitesimal quantities to be paraphrased? If so, how?

It was perhaps Newton, with his somewhat more cumbersome and heavy-handed approach to the calculus, who came closer to an understanding of what was 'really' going on. In his more self-conscious moments, he acknowledged the problematical nature of infinitesimals and made suggestions as to how they should be eliminated. Here are two quotations:

64

In the method of fluxions ... there is no need to introduce infinitely small quantities;[4]

and:

These ultimate ratios with which the quantities vanish are indeed not ratios of ultimate [*sc.* infinitesimal] quantities, but limits to which the ratios of quantities vanishing without limit always approach, to which they may come up more closely than by any given difference but beyond which they can never go.[5]

The latter quotation presages nicely the kind of sophistication and rigour which, as we shall see, were later infused into the calculus.[6]

Beyond

Neither Leibniz nor Newton, then, nor any of the other great mathematicians who first took up these methods, was oblivious to the problems that they raised. But the point was, they worked. That is, they worked in a very practical sense. Given certain commonly accepted principles of nature, including, eventually, Newtonian mechanics, one could use these methods to calculate forces, velocities, rates of acceleration, areas, volumes, and the like, and in a way that squared with the observed data. It was difficult not to be impressed by such success. Still, from the point of view of true mathematical understanding, this was not enough.

It was the Irish philosopher George Berkeley (1685–1753), the Anglican bishop of Cloyne, who was to become the most famous critic of the calculus. Aggrieved by the fact that the very same mathematicians who accepted these incoherent methods also objected to various principles of Christianity on grounds of incoherence, he mounted a courageous and scathing attack against them; courageous, because of the popularity and success that the methods were beginning to enjoy. He wrote an essay entitled *The Analyst; or, A Discourse addressed to an Infidel Mathematician. Wherein it is examined whether the Object, Principles, and Inferences of the modern Analysis are more distinctly conceived, or more evidently deduced, than Religious Mysteries and Points of Faith. 'First cast out the beam out of thine own eye; and then shalt thou see clearly to cast out the mote out of thy brother's eye.'*[7] Berkeley pointed out, quite rightly, that even if the calculus led to true conclusions, this did not vindicate it as a genuine science. Truth can be arrived at when errors cancel one another out.

Berkeley's criticisms were certainly justified. One of the objects of his attack was the first textbook on the calculus, written by the French mathematician G.F.A. de l'Hôpital (1661–1704).[8] This did an enormous amount to help popularize the calculus, but it was steeped in the kind of confusion that came with a completely uncritical acceptance of the infinitely small. He wrote:

65

A quantity which is increased or decreased by a quantity which is infinitely smaller than itself may be considered to have remained the same;

and:

A curve may be regarded as the totality of an infinity of straight segments, each infinitely small: or ... as a polygon with an infinite number of sides.

In the light of the work that Eudoxus and Archimedes had done two thousand years earlier, this second quotation is especially striking.

Another influential textbook on the calculus was later published by the Swiss mathematician Leonhard Euler (1707–1783).[9] This masterful work was notable for the way in which it brought things together into a beautifully organized whole that could be understood in purely analytic (non-geometrical) terms. But it still made use of infinitesimals in an unsatisfactory way. And as a result, dissatisfaction persisted. Hegel took up criticisms of the calculus that were similar to Berkeley's (though he spoke rather favourably of Euler); and he made appropriate suggestions as to how it should be straightened out.[10]

One thing was becoming increasingly clear. For all its utility, a rigorous account was badly needed of what was going on in the calculus – *if* it was going to be an object of true mathematical understanding. As should already be clear, and as was implicit in suggestions made by Hegel and others, the seeds of such an account lay in antiquity, in the work of Eudoxus and Archimedes. Not that their work had been ignored in the seventeenth century. On the contrary, their methods had been exploited in the solution of many geometrical problems and even invoked in early work pertaining to the integral calculus. But it was not until the nineteenth century that the rigours underlying these methods were properly resuscitated to give the calculus, at last, a secure and respectable foundation.

Two mathematicians played an especially prominent role in this resuscitation: Augustin Cauchy (1789–1857), a Frenchman whose ideas were later echoed and endorsed by Bolzano; and, above all, Karl Weierstrass (1815–1897), who was born in Westphalia, and in whose hands the techniques of the calculus were finally sharpened and refined to the point of irreproachability. Their fundamental ideas were there already in the method of exhaustion and its justification. Eudoxus and Archimedes had shown that, when using this method, we did not need to think of a curved figure as an 'infinigon'. We could see it as the limit of a sequence of polygons, which we must in turn understand not in objectionable infinitary terms but in terms of generalizations that we can make about the polygons in their relation to the figure. Cauchy and Weierstrass argued that the same reasoning was applicable elsewhere. (Had not William of Ockham's account of points already testified to this (see above, Chapter 3, §3)?)

To gain a feel for their approach reconsider the tangent at P whose slope was pondered in §1. We do not need to think of this as a line joining P to another point infinitesimally close. Instead we can consider lines joining P to points some definite distance along the curve (in either direction). Following through the argument in §1, we can soon see that all these lines have slopes of the form $(2x\delta + \delta^2)/\delta$, where δ is a real number other than 0, positive or negative (and more or less 'small' according to whether it is closer or not to 0). We can then specify and calculate the slope of the tangent at P by thinking of it as the limit of these lines, and understanding this as the ancients would have. How would the ancients have understood it? They would have pointed out that the smaller δ is, the closer the quotient $(2x\delta + \delta^2)/\delta$ is to the slope. But not only that. (Remember the 'deformed' circle in Figure 1.6 of Chapter 1.) They would also have noted that the quotient gets as close to the slope as you care to specify. That is, no matter how small a number ε you consider (a trillionth, say), there will always be some value of δ that is itself sufficiently small for $(2x\delta + \delta^2)/\delta$ to lie within ε of the slope, and likewise of course for all smaller values of δ. It is in this sense that $(2x\delta + \delta^2)/\delta$ gets *arbitrarily* close to the slope, for different values of δ. It then follows that the slope is $2x$. The principle here can be expressed as follows:

(1) the limit of $(2x\delta + \delta^2)/\delta$, as δ tends to 0, is $2x$,

and written as follows:

$$(2)\ \lim_{\delta \to 0} (2x\delta + \delta^2)/\delta = 2x.$$

This does not mean that we are evaluating $(2x\delta + \delta^2)/\delta$ in the case where δ is 0. (There is no such case. Division by 0 is illegitimate.) Nor are we envisaging δ infinitely small. Rather, we are making a kind of generalization about different finite values of δ. When we say that the slope of the tangent at P is $2x$, what we are saying can be understood in terms of perfectly legitimate generalizations about finite numbers, lines, and the like.

(1) and (2) are basically claims about what happens as δ gets arbitrarily small. We also sometimes want to consider what happens as values get arbitrarily large. Thus, for example, we might say:

(3) the limit of $1/n$, as n tends to infinity, is 0,

and we might write this as follows:

$$(4)\ \lim_{n \to \infty} 1/n = 0,[11]$$

where n, in each case, is understood to take natural numbers as its values. But again, this does not mean that we are evaluating $1/n$ in the case where

n is infinite. (No natural number is infinite.) We are saying that no matter how small a number ε you consider, there is always a value of n that is sufficiently large for its reciprocal, and the reciprocals of all larger values, to lie within ε of 0 (that is, to be smaller than ε). There is no appeal to the infinite here. It is clear that the word 'infinity' in (3), and the symbol '∞' in (4), appear by courtesy only. (3) and (4) are perfect examples of *façons de parler*. One could imagine the medievals extending their notion of the syncategorematic to incorporate such uses of 'infinity'.

Here is another application of these ideas. Consider the following infinite sequence of rationals:

$$\langle \tfrac{1}{2}, \tfrac{3}{4}, \tfrac{7}{8}, \ldots \rangle$$

The limit of this sequence is 1. This means: however small a number ε you consider, there is a rational sufficiently far along the sequence for both it, and all the other rationals in the sequence thereafter, to lie within ε of 1. The rationals in the sequence get arbitrarily close to 1, even though none of them is itself equal to 1. (I said in the Introduction that concepts like unlimitedness, endlessness, and infinitude, once made precise in various standard ways, turned out to be different from one another. Here is a sequence that is infinite and endless, yet limited in the straightforward sense that it has a limit.)

Now we saw in Chapter 1, §2, in effect, that an infinite sequence of rationals can have a limit that is not itself a rational. For example, the limit of the sequence

$$\langle \tfrac{1}{1}, \tfrac{3}{2}, \tfrac{7}{5}, \tfrac{17}{12}, \ldots, x/y, (x + 2y)/(x + y), \ldots \rangle$$

is $\sqrt{2}$. Of course, appeal to this limit would cut no ice with an inveterate Pythagorean who refused to believe in irrational numbers. He would simply deny that the sequence *had* a limit. Not all sequences do. Here are two sequences which, for very different reasons, do not:

$$\langle 0, 1, 2, 3, \ldots \rangle$$

and

$$\langle 0, 1, 0, 1, \ldots \rangle$$

(He would also have to deny that 2 had a square-root.) He might be won over by a geometrical construction, just as the original Pythagoreans had been. But he could hardly be swayed by analytic considerations.

But consider: the calculus is ultimately to be carried out in purely analytic terms. It ought not to rest at any point on ineliminable appeal to geometrical intuition. So even if we cannot be expected to supply an irresistible non-geometrical proof of the existence of irrational numbers, we do need a suitable and careful explanation of how they are supposed to fit in. It was the German mathematician Richard Dedekind (1831–1916) who first noted this need. He went back to the natural numbers and various fundamental operations that could be performed on them, and set about giving an account, in these terms, of, so to speak, the logical gaps that the irrational numbers had to fill. (One such gap, of course, was created by the fact that 2 had no rational square-root. But there were other gaps of a different kind.) He thought that in the end the irrational numbers had to be 'created' or 'constructed' to fill these gaps.

I shall not dwell on the details of his procedure. But it is worth noting that, when we return to a geometrical understanding of this, the gaps strike us as gaps of a much less metaphorical kind. Let us suppose that we can make sense of a line made up of points corresponding to just the rational numbers between 0 and 2, ordered accordingly. (I do not believe that we can make sense of this (see below, Chapter 10, §5). But it is a useful heuristic device.) Then such a line would contain 'real' gaps; there would be one such gap where the point corresponding to $\sqrt{2}$ should be.[12] What is interesting is that the line would still be infinitely divisible, or infinite by division, in the sense that has been adopted in this book. Between any two points on it there would be a third. (Between any two rationals there is a third. Actually, we do not need to invoke irrationals to prove that infinite divisibility, in this sense, allows for gaps. If one 'took away' the point on the line corresponding to $1\frac{1}{4}$, say, thereby deliberately creating a gap, it would still be true that between any two points on the line there was a third.) So Dedekind's work helped to show that the notion of infinite divisibility with which we have been operating is of something somewhat more porous than it might have been. It was in this connection that he developed the notion of *continuity*. (Cantor developed a closely related notion independently.) Very roughly: an imposition of order on a set of things is *continuous* when it is not only true that between any two of them there is a third, but also that there are no 'gaps'. For example, the standard imposition of order on the reals is continuous. An imposition of order which simply has the first of these properties, namely that between any two of the things ordered there is a third, is said to be *dense*.[13]

A final point in this section: the German logician Abraham Robinson (1918–1974), who invented what is known as *non-standard analysis*, thereby eventually conferred sense on the notion of an infinitesimal greater than 0 but less than any finite number. But in making this sense precise, he used logical methods and techniques that went far beyond what would have been recognizable to seventeenth-century mathematicians. It would be anachronistic to see his work as a vindication of what they had been up to. It did not show that the notion of an infinitesimal as understood by *them* had been coherent.[14]

The Infinite

3 Taking stock

Clearly the calculus bears on the paradoxes of the infinitely small. I do not myself believe that we need it to solve any of them; they are soluble, if at all, without recourse to such sophisticated machinery. But we can use it to formulate more crisply certain fundamental features of the situations involved. Consider, for example, Zeno's arrow paradox. Suppose we have an account, of a broadly Aristotelian kind for example, of what it is for an arrow to be in motion at an instant, in a derivative sense. The calculus enables us to go much further. It enables us to say what it is for the arrow to have a particular *velocity* at that instant.

In order to see further bearing of the calculus on these paradoxes, it will be helpful to consider two more. They are modern embellishments of the earlier paradoxes. The first is due to Thomson, the second (in essentials) to Benardete.[15]

(i) The paradox of the lamp

Suppose there is a lamp with a switch which, when depressed, turns the lamp on if it was off and off if it was on. Now suppose that the lamp is off at some initial point, then turned on half a minute later, then turned back off a quarter of a minute later than that, and so on *ad infinitum*. Will it be on or off exactly one minute after the initial point? We do not know how to answer, yet we feel that there must be an answer.

(ii) The paradox of the space ship

Suppose there is a space ship which, after travelling in a straight line for half a minute, doubles its speed, and, a quarter of a minute later than that, doubles its speed again, and so on *ad infinitum*. Where will it be at the end of the minute? Infinitely far away? This does not make sense, but it is not clear what else we can say.

Of course, each of these paradoxes invites scepticism about the coherence of its description. In the paradox of the lamp, for example, a situation has been described that is certainly not compatible with all that we believe about the physical world, and perhaps not compatible with the structure of space and time. (That structure may itself impose a limit on how quickly the lamp can be switched on and off.) Aristotle would have denied that it was even compatible with a correct understanding of the infinite.

But we can be sceptical in another way. Each paradox arises from our not being able to answer a certain question about the situation described. Our scepticism might be trained not on the coherence of the description, but on its determinacy.

To see how, we must first recall one of the lessons of the paradox of the arrow. When we say that something is in a particular state at a particular instant, this sometimes has to be taken in a derivative sense. For example, if we say that an arrow is in motion at a particular instant, then we must mean that the instant occurs in a period of time throughout which the arrow is moving. Now on this kind of understanding, the lamp is neither on nor off at the end of the minute. For the end of the minute is not an instant that occurs within a period, however small, of its being (exclusively) either. So as not to evade the issue, however, let us grant that something can be in a particular state at a particular instant if the instant *initiates* a period throughout which the thing is in the state. It is now that the question of determinacy looks acute. Each paradox, it now appears, concerns time after the end of the minute. Once things are understood in these terms, it seems clear that, even if the lamp were switched on and off in the way described, its state at the end of the minute simply would not (thereby) have been determined. It is compatible with how the situation has been described – *modulo* the coherence of the description – that the lamp should be on at the end of the minute, equally compatible that it should be off, and equally compatible for that matter that it should have disappeared altogether. Again, it is compatible with all that we have been told about the space ship that it should be here at the end of the minute, there, or anywhere else, and that it too should have disappeared altogether. (I am glossing over the point that if spatial continuity is essential to a space ship's identity over time, then it *must* have disappeared altogether.) Asking where it *would* be is akin to asking where it was a minute *before* the journey, or how tall Hamlet was. Given all the relevant information, there is, in each case, no fact of the matter.

Here too the calculus can help us to put this more crisply. Consider the (overlapping) stretches of time after the initial point in each paradox, with lengths of half a minute, three quarters of a minute, and so on *ad infinitum*. These have the full one-minute stretch as their limit. They get arbitrarily close to it. But none of them *is* it. We can know what happens at the end of each of *them* without knowing what happens at the end of *it*.

This is also relevant to Achilles in Zeno's paradox of the runner. In running from *A* to *B* he must perform all the tasks specified in the paradox. But he could perform them all and still not be (logically) guaranteed arrival at *B*. He too might suddenly disappear. This leaves us, of course, with the problem of specifying how and in what sense he can perform the infinitely many tasks (making sure that we take into account the related paradox of the staccato run). Here, however, the relevant considerations are not those that inform the calculus. They are those that exercised Aristotle.[16]

Here is another way in which the calculus can be used to impose conceptual clarification.

The Infinite

Mathematicians sometimes talk about infinite sums. They might, for example, assert the following:

$$(5) \; \tfrac{1}{2} + \tfrac{1}{4} + \tfrac{1}{8} + \ldots = 1.$$

But it is not obvious what this means. Addition is ordinarily defined for finite input. So on the strength of our ordinary understanding of addition we have no way of assessing whether (5) is correct or not. If we thought we had, we would be embarrassed when it came to evaluating the following 'sum':

$$1 - 1 + 1 - 1 + 1 - \ldots$$

(The mathematician Grandi drew attention to this 'sum' early in the eighteenth century.) Appropriate bracketing can make this seem equal first to 0 and then to 1, witness each of the following:

$$(1 - 1) + (1 - 1) + \ldots;$$

and

$$1 + (-1 + 1) + (-1 + 1) + \ldots$$

We might even call the 'sum' x, argue that $x = 1 - x$, and conclude that $x = \tfrac{1}{2}$. But the lesson is clear. Equations such as (5) are not to be accepted until a precise account has been given of what exactly they mean.

There is a natural account, however, made available by the calculus. This is to identify the sum with the limit, if there is one, of the successive finite sums; and, if there is no limit (as in Grandi's example), to deny that there is a sum. Let us accept this account. Then (5) is true because the limit of the sequence

$$\langle \tfrac{1}{2}, \tfrac{3}{4}, \tfrac{7}{8}, \ldots \rangle$$

is 1. And this, as we know, does not mean that 1 is the infinitieth term in the sequence, a notion without sense. (5) is true not because of the outcome of an infinite addition but because of a certain generalization about the outcomes of finite additions.

We can now see why the definition of real numbers proffered in Chapter 1, §2, as those which can be expressed using infinite decimal expansions, needed clarification. For an infinite decimal expansion simply picks out an infinite sum. To say that $\pi = 3.141 \ldots$ is to say that

$$\pi = 3 + \tfrac{1}{10} + \tfrac{4}{100} + \tfrac{1}{1000} + \ldots$$

But this we now explain as follows. π is the limit of the sequence

$$\langle 3, 3\tfrac{1}{10}, 3\tfrac{14}{100}, 3\tfrac{141}{1000}, \ldots \rangle$$

(It might have seemed that infinite decimal expansions were all that Dedekind needed to invoke when trying to specify how irrational numbers fitted in. But we can now see that this would not really have advanced his cause. He already knew that irrationals were the limits of certain sequences of rationals.)

Where does all of this leave the broader issues of the infinite? Back with the Greeks, it seems. It is true that Dedekind's construction of the irrational numbers forced him to take infinite sets into account. (It was similar work by Cantor that led him to the view that infinite sets were legitimate objects of mathematical study (see below, Chapter 8, §3).) And certainly, any mathematical account of the real numbers must acknowledge the special way in which they transcend the finitude inherent in the natural numbers – as the Pythagoreans learnt to their cost. But to the extent that the real numbers are themselves only finite quantities, it seems that early Greek hostility towards the mathematically infinite, or at least Aristotelian hostility towards the actually mathematically infinite, have been given a further boost. For what the calculus seems to do, once it has been suitably honed, is to enable mathematicians to proceed apace in just the sort of territory where the actual infinite might be expected to lurk, without having to worry about encountering it. They can uphold claims ostensibly about infinitesimals or about infinite additions, and they can even use the symbol '∞', knowing that they are only making disguised generalizations about what are in fact finite quantities. They still need not look the actual infinite squarely in the face.

They need not perhaps. But it does not follow that they cannot. We saw in the last chapter a growing friendliness in some quarters towards the actual infinite, based on the conviction that its paradoxes, specifically the paradoxes of the infinitely big, could be alleviated. What we have seen in this chapter could well bolster such a conviction, though this time with respect to the paradoxes of the infinitely small. I have already said that, in my view, we do not need the calculus to solve those paradoxes. But it certainly lends substance to some of the more radical (non-Aristotelian) solutions. Thus, for example, somebody might say that there is nothing incoherent in the idea of performing infinitely many tasks in a finite time (even if this involves writing down the complete decimal expansion of π); and that the notion of a limit helps to vindicate this. Again, they might say that there is nothing wrong with regarding a line as made up of an actual infinity of points (provided that this is interpreted with due care, and not, for example, in such a way that the length of the line is supposed to be an infinite sum of the lengths of the points); and that the notion of a real number, now refined, helps to vindicate this.

But more importantly, somebody might urge that self-conscious reflection on the mathematical techniques employed in the calculus *forces* us to look the actual infinite squarely in the face. It is all very well saying that the calculus involves generalizations about finite quantities, but the whole thing only works because there are infinitely many *of* them. For instance, an infinite sum is made intelligible in terms of a certain sequence of finite sums – an infinite sequence. I remarked in Chapter 1, §5 that early Greek mathematics was committed to the infinite in an indirect way. It did not involve explicit reference to the infinite, but it needed the infinite (so to speak) to accommodate it. We see something similar in the calculus. The point was, and is: study of what is finite is sometimes possible only in an infinite framework. And there is nothing – is there? – to rule out self-conscious reflection on that framework.[17]

CHAPTER 5

The Rationalists and the Empiricists

'Tis universally allow'd, that the capacity of the mind is limited, and can never attain a full and adequate conception of infinity: And tho' it were not allow'd, 'twou'd be sufficiently evident from the plainest observation and experience. (David Hume)

The eternal silence of these infinite spaces fills me with dread. (Blaise Pascal)

The mainstay of seventeenth-century continental philosophy was rationalism. Two of the three philosophers most frequently classified as rationalists, Descartes and Leibniz, have already made a large impact on this enquiry through their work in mathematics; the third, Spinoza, so revered mathematical method that he modelled his major work (his philosophical masterpiece *Ethics*) on Euclid's axiomatization of geometry. This is revealing, because British philosophy, that same century and the next, saw a backlash, in the form of empiricism, and the central point of controversy between rationalism and empiricism was the extent to which understanding of the world could be arrived at by *a priori* means – by that exercise of pure reason which is characteristic of mathematics. The rationalists held that deep substantial truths about the structure of the world – not just mathematical truths – could be discovered in this way. The empiricists insisted that it was only through experience that we could come to know such truths. It is worth noting the similarities between the empiricists' reaction to rationalism and Aristotle's reaction to the more metaphysical strains in *his* predecessors. In this chapter I shall try to spell out the implications of all of this for continuing thought about the infinite.

1 The rationalists

Before Leibniz

In Chapter 3 we saw a growing friendliness towards the actual mathematical and the metaphysical infinite, which had begun to break the mould

75

of Aristotelianism. The rationalists, in various ways, consolidated this friendship. Because they were not particularly constricted by the demands of experience, they believed that they could accommodate the non-potential infinite despite the obvious experiential and imaginative limitations that we encounter when trying to understand it. They thought that even though we could not meet it in experience or grasp it in imagination, we still had a perfectly clear and determinate conception or idea of it, innate within us; and such an idea constituted, or helped to constitute, a fundamental and vital insight into reality.

I suggested in the Introduction that the strongest pressure on us to acknowledge the infinite arises from a contrast with our own finitude. This thought has recurred many times throughout the history of the infinite. But the rationalists took just the opposite view. They held that our idea of the infinite enjoyed both a logical and an epistemological priority over that of the finite. This view was boldly put forward for the first time by Descartes. Indeed he based one of his arguments for the existence of God on it. He argued that since God alone was truly infinite, He alone could have implanted our idea of the infinite in us. Our own finitude and limitations meant that we could not grasp such infinitude, and it was idle for us to speculate about it in what were inevitably finite terms (to enquire, for example, whether infinite number was odd or even, or what resulted if an infinite magnitude was halved). But not being able to grasp it did not preclude touching it with our thoughts, any more than not being able to grasp or embrace a mountain precludes touching it. Reason could put us in touch with the infinite, even if experience and imagination could not.[1]

This was a keynote of rationalism. It was later struck with wonderful lyricism by Descartes' compatriot Blaise Pascal (1623–1662). He expressed with an eloquence that has never been surpassed the wonder, awe, and dread that we feel in the face of the enormity and minuteness of nature, which he described as 'an infinite sphere whose centre is everywhere and circumference nowhere.' Her two infinite extremes have their meeting place in God alone, he said, and man finds himself lodged between these extremes, bewildered by their utter ungraspability; he can never hope to imagine or to understand them. 'Man is only a reed, the weakest in nature,' he wrote. 'But,' he added, 'he is a thinking reed.' Indeed the basic principle of morality, he held, was to think well. And this captured beautifully the spirit of rationalism: despite man's helpless vulnerability in the midst of limitless nature, he can, through his rationality and thinking, achieve a nobility and dignity that transcend his finitude; and in this way he can touch the infinite.[2]

By contrast, Pierre Gassendi (1592–1655), another Frenchman, anticipated empiricist objections to all of this and insisted, against Descartes, that we could not have a positive conception of what we could not, in Descartes' sense, grasp. He believed that our idea of the infinite really was

76

a mere negation of our idea of the finite. We recognized the possibility of limitless augmenting of the finite, and arrived at the idea of the infinite by a kind of negative extrapolation from this.[3]

But Descartes retorted adamantly that it was not so. What we arrived at in this way was an idea of what he called the *indefinite*, a pale reflection of the infinite, grounded in a mere lack of observable limits. Our idea of the infinite, for example our idea of the limitlessness and perfection of God, was an idea of something that positively transcended this, something actual rather than potential, something utterly timeless, a kind of amalgam of the metaphysical and the mathematical. In a thoroughly un-Aristotelian spirit, he wrote in a letter to Clerselier as follows:

> I never use the word 'infinite' to mean only what has no end, something which is negative and to which I have applied the word 'indefinite', but to mean something real, which is incomparably bigger than whatever has an end.[4]

The Dutch philosopher Benedictus de Spinoza (1632–1677) was a man of deep religious conviction (though both his fellow Jews and Christians viewed his mystical writings as utterly unorthodox, if not downright atheistic). He was philosophically much indebted to Descartes, as well as to Neoplatonic strands in medieval thought. A characteristic feature of his philosophy was the way in which he pushed various Cartesian principles to their logical limit. We see this in his views on infinity. He agreed with Descartes that God was positively infinite, in a way that we could understand with our intellect but not with our imagination. But he believed that, once this had been thought through, such infinitude could only be understood in radically metaphysical terms. He attempted to prove, in a rigorous and axiomatic way, that God was an absolute indivisible unified whole, Whose essence was to exist and to be all that exists. His very infinitude meant that nothing else could exist, lest it should limit Him; nor could He be divided into parts, for these, if genuinely separate, would all have to be finite, in which case nothing would be left of His indestructible infinitude. (Spinoza thought that this circumvented the paradoxes of the infinitely big that arise when we attempt to make comparisons of size between an infinite whole and its parts.) God was the simple eternal one true substance. Spinoza's thinking here was reminiscent of that of Parmenides and Plotinus. But, unlike them, and again rigorously following through basic rationalist assumptions, he adopted a pantheism according to which this fundamental reality was as immanent as it was transcendent. To this extent, his thinking was more reminiscent of that of Nicholas of Cusa. (This was also the reason for his being branded an atheist.) Such pantheism was inevitable for Spinoza. Since God's reality was of every kind, one of His attributes had to be (physical) extension. God and nature were one. Bodies were not anomalies or illusions in this scheme of things. They were, as Spinoza would have put it, determinate 'modes' by means of which God's extension was expressed.[5]

Spinoza's account of extension led him to draw an important distinction. There was absolute infinitude, which God alone enjoyed, the infinitude of that which encompasses the whole of reality. But there was also a weaker notion, 'infinitude in its own kind'. This was enjoyed by whatever was not limited by anything else of the same kind (unlike the absolute infinite, which was not limited by anything else *at all*). Space and time were each infinite in the weaker sense. By contrast, an individual body, hemmed in by others, was not even that.[6]

It seems reasonable to view Spinoza's distinction as a kind of reformulation of the metaphysical/mathematical distinction. If this is right, then we have here another glimpse of what has been a recurring theme in this enquiry: the idea of a metaphysically infinite reality with mathematically infinite aspects. In its most eclectic form, this idea merges two great Hellenic traditions: it involves a radical distinction between reality and appearance that owes much to the Eleatics and to Plato; and it involves an understanding of the mathematical infinitude of appearances that owes much to Aristotle. We saw it first and most graphically in the Neoplatonic tradition with Plotinus and Augustine. We saw it also to some extent in Aquinas, and later in Nicholas of Cusa and Bruno. We see it next in Leibniz.

Leibniz

Along with the other rationalists, Leibniz held that true infinity could be grasped by the intellect but not by the imagination. The truly infinite was the absolute, God, Who in His absoluteness was not formed by the addition of parts but preceded all composition. His infinitude was of a metaphysical kind. He was also outside space and time – *pace* Spinoza. Space and time were features of how things appear to us. But even in their case, Leibniz thought that we had an idea (not an image) of the infinite innate within us, an idea that was not simply arrived at by extrapolation from our idea of the finite. This was the mathematically infinite (by addition and by division). Leibniz' views clearly conformed to the pattern just outlined.

But how did he view the mathematically infinite? In the Aristotelian way that had characterized the more eclectic versions of this position?

At first blush, not at all. There were some Aristotelian elements in his thinking. For instance, he had roughly Aristotelian reasons for thinking that infinite space ruled out a spatially finite world: space was homogeneous, so there could be no reason why any part of it should have the privilege of housing such a world. But elsewhere, his views looked strikingly unlike Aristotle's. In a letter to Foucher he wrote:

I am so much in favour of the actual infinite, that ... I hold that nature affects it everywhere, in order the better to mark the perfection of its author. So I believe that every part of matter is, I do not say divisible, but actually divided.[7]

Again, commenting on his belief in the continuity of all change, he wrote of:

the immeasurable fineness of things, which always and everywhere involves an actual infinity.[8]

Certainly Leibniz left no room for doubt that he regarded space, time, and that which is in space and time as infinite both by addition and by division, and in some sense actually so. This was the source of another bitter controversy between him and his followers on the one hand, and the followers of Newton on the other. The latter held the world to be finitely big and composed of finitely many indivisible atoms. There is a famous correspondence between Leibniz and one of Newton's best-known disciples, the English philosopher Samuel Clarke (1675–1729), in which these and other issues were thrashed out with passionate verve.[9]

We must not forget, however, that Leibniz' views here were tied up with a belief in the unreality of space and time. (Indeed they substantiated it. Infinite divisibility required wholes with parts whose existence was parasitic on the wholes, and this, in Leibniz' view, flouted a basic metaphysical principle of what is real.[10]) This already served to mitigate whatever sense of actuality was being intended. And indeed he went on to insist that neither space nor time, nor that which is in space and time, could be treated as actual measurable wholes. Borrowing the term introduced by the medievals, he said that nature was infinite only in a syncategorematic sense. That is, given any finite part of nature, there was always more to come. This sounds much closer to Aristotle.

Moreover, there were, for Leibniz, no infinite quantities or infinite numbers; the paradoxes of the infinitely big showed that. For example, the paradox of the even numbers showed that it was illegitimate to assign a number to the collection of natural numbers.[11] Similarly, we could say that to each even natural number there corresponded a unique natural number and *vice versa*, but we could not say that the number of natural numbers was equal to the number of those that were even. At the other extreme Leibniz also disavowed numbers that were infinitely small. As we saw in the last chapter, he thought that reference to them could be legitimated as a purely technical device within the calculus. But it had to be taken as just a convenient *façon de parler*, in the same way that describing a straight line as a circle with an infinite radius was (in certain contexts) a convenient *façon de parler*. In general, then, Leibniz' views concerning the mathematically infinite did contain an important element of Aristotelian restraint.[12]

2 The empiricists

The three philosophers most frequently classified as empiricists are Locke, Berkeley, and Hume, who were respectively English, Irish, and Scottish. (Berkeley appeared in the last chapter in connection with the calculus.) To a large extent their philosophy was a reaction to rationalism. Not that the dialectic current flowed entirely one way. On the contrary, most of what we have just seen of Leibniz' views on the infinite was presented as a direct response to Locke, in a commentary on the latter's major philosophical work.[13] But empiricism was first and foremost a recoil from rationalist claims on behalf of reason. The empiricists held that we could have no idea or concept of anything that did not first somehow impinge on us through experience. In particular, of course, this was true of the infinite. The rationalist catchword of the previous section, that we could form an idea of the infinite even though we could neither experience it nor imagine it, struck them as fundamentally awry.

It is clear, then, that the infinite was going to present the empiricists with a major challenge. For the fact is that we *cannot* experience it or imagine it, not in any straightforward sense. This is true even of Aristotle's potential infinite. And this is significant, for a kind of empiricism had been what motivated Aristotle when he championed the potential infinite in the first place and eschewed earlier metaphysical accounts. Something more radical was evidently going on here – already in Locke, and successively more so in Berkeley and Hume. Here, if anywhere in the history of philosophy, there was a danger that the most extreme measure would be taken in response to all the problems that the infinite posed; I mean the measure of dismissing the very concept of the infinite as empty or incoherent.

In fact it was taken at a very early stage, even before any of the three empiricists mentioned above had addressed the problem, by the great English philosopher Thomas Hobbes (1588–1679). Hobbes was responding, in part, to Descartes. He insisted that we could conceive nothing that we had not previously perceived (either all at once, or by parts), and that we therefore had no real conception of the infinite – certainly not as opposed to the indefinite of which Descartes had spoken. In Hobbes' view, when people described something as infinite they were, at best, making an indirect comment on their own inabilities, for example their inability to conceive an end of it.[14]

What about the world as a whole though? It seemed as embarrassing for Hobbes to deny that it was infinite as to admit it, for, after all, no one had ever perceived, or perhaps could perceive, a limit of it. (Reconsider Archytas' argument (see above, Chapter 1, §5).) He responded to this dilemma with wonderful pertinacity, indeed with a pertinacity that bordered on the perverse. He wrote:

When it is asked if the world is finite or infinite, there is nothing in the mind corresponding to the vocable *world*; ... whatever we imagine is *ipso facto* finite.[15]

John Locke (1632–1704) could not bring himself to go quite so far. He denied only that we had a positive idea of the infinite. But he thought that our ideas of space, time, and number were such that we recognized the possibility of increasing them without limit, and, in the case of space and time, dividing them without limit; and he thought that this furnished us with a viable *negative* idea of the infinite. What did he mean by this? It sounded very like what Gassendi had said in response to Descartes (see above, §1). But it was also highly reminiscent of Aristotle. What he meant was basically this. Our idea of the infinite was not something present to our minds 'all at once'; it involved us in running through certain processes in our minds, such as counting or imagining larger and larger volumes of space, and recognizing that they need never terminate. In the case of space he expressed his position as follows: whereas we had an idea of the infinity of space, we had no idea of space infinite. Similarly with time and number. (He also argued, incidentally, that there was no problem with a finitely big world's being lodged in otherwise empty infinite space. This set him apart from a tradition going right back to Aristotle and continuing, as we saw, through to Leibniz. But it put him in line with the Newtonians.) It was when we treated our idea of the infinite as if it were positive, and thus something to be encountered all in one go, that we became entangled in confusion and paradox, he said. Any quantity of which we could have such an idea was capable of increase. But it was absurd and contradictory to suppose that the infinite was capable of increase, or (what followed from this) that one infinity could be greater than another.[16] The infinite was not a determinate or finished whole. In this respect, and in respect of his rejection of infinite number, Locke was, of course, in company with Zeno, Aristotle, Leibniz, and a multitude of others.[17]

But now consider: in Locke's scheme *something* had to be present to our minds 'all at once', for how else could we know in advance that the relevant processes were capable of being carried on indefinitely? In admitting even a negative idea of the infinite Locke was, in effect, conceding that there was more to the mind's conceptual apparatus than ideas, copied, image-like, from previous direct experience. I do not say that Locke was unaware of this. But it showed that his empiricism was still of a somewhat mitigated kind, retaining crucial rationalist elements.

Berkeley and David Hume (1711–1776) returned to the more radical position of Hobbes. The quotation from Hume at the beginning of this chapter bears witness to this.[18] A very few deprecatory remarks of that kind aside, both Berkeley and Hume made clear their discomfort with the idea of the infinitely big by not discussing it at all, even when dealing with space or time or number. This was a potent way of registering that on their view there was nothing there to discuss. They made their position on the infinitely small equally clear. They maintained that there was a definite limit to how small anything could be, spatially or temporally. (Berkeley was more specific. He denied that there was any such thing as the ten thousandth part of an inch.) What reasons did they have for this? First, only an atomistic view of space and time could allow for one line, or one period of time, to be longer than another (by containing more points or moments). Secondly, only an atomistic view of time could allow for temporal succession, which required a passage from one moment to 'the next'. These were Hume's arguments. But also, they both argued that we could form no clear idea of what the alternative would be like; for experience itself was composed of indivisible atom-like constituents – '*minima sensibilia*'. To establish this, Hume invited us to consider the appearance of a spot of ink as we backed away from it, the moment before it vanished. And as against those thinkers who argued on *a priori* grounds that space and time *must* be infinitely divisible, he retorted that, on the contrary, space and time must be capable of being how they struck us as being in experience. (In fact both Berkeley and Hume developed their empiricism to the point where things in space and time were of the same 'stuff' as experience anyway. But the issues that arise here are largely independent of the subtleties of such a metaphysics.)[19]

But how did Berkeley and Hume view mathematics, which in various ways seemed to tell against their position? We have already seen how Berkeley reacted to the calculus, a reaction that now makes all the more sense. We must not think, however, that his criticisms required the backing of this atomism. One can believe in infinite divisibility and still find the idea of an infinitesimal incoherent. Indeed geometry seems to presuppose the one but not the other. What did Berkeley and Hume make of geometry then?

Berkeley argued, ingeniously, that when geometricians spoke of a given line as being divisible without limit, really what they were doing was taking it to be representative of lines of arbitrary length; that is, they were speaking syncategorematically rather than categorematically, though in a way that was misleading (and had indeed misled them).[20] Hume simply regarded geometry as an inexact science that was based on experience but misrepresented it in various ways.[21]

This boldness was entirely characteristic of Hume. On this issue, as on so many others, he displayed a marvellously resolute commitment to empiricist principles. He was prepared to follow them wherever they led. But they had now surely led too far. We cannot simply jettison the concept of the infinite in this way (as Aristotle had realized some two thousand years earlier). There can be no doubt that empiricism was one of the great philosophical movements, of deep and lasting significance: but this is partly because of the lessons we can learn from its ultimate failure. It is a very important feature of the infinite that it helped to signal that failure.[22]

CHAPTER 6
Kant[1]

When I look up at Thy heavens, the work of Thy fingers, the moon and
the stars set in their place by Thee, what is man, that Thou shouldst
remember him, mortal man that Thou shouldst care for him? Yet Thou
hast made him little less than a god, crowning him with glory and
honour. (Psalm VIII)

1 The background: an outline of Kant's philosophy[2]

It is often said that the three greatest philosophers of all time were Plato,
Aristotle, and Kant. The third of these now makes his mark on our enquiry.

Immanuel Kant (1724–1804) was born in Königsberg in Prussia and lived
there all his life. Much of his philosophy was devoted to taking rival
systems of thought and rooting out the inveterate assumptions common to
them. On the one hand this enabled him to show that some of the
fundamental points of controversy between them were ill-conceived. On
the other hand it enabled him to salvage and to reconcile some of their
apparently irreconcilable insights. The latter was something that he sought
to do above all in the case of the conflict between traditional Christian
morality and Newton's (by now) well-established mechanics. Christian
morality seemed to make no sense without human freedom, but there was
no room for human freedom, it seemed, in Newton's world of inexorable
mechanical laws. In attempting a reconciliation here, Kant developed a
philosophical system of breathtaking depth and power. Indeed it enabled
him at the same time and in much the same way to arbitrate between the
rationalists and the empiricists. Since they have just been the focus of our
attention, let us broach Kant's system in those terms.

Kant wanted to accept, with the rationalists, that we had substantial *a
priori* knowledge. Yet, in line with the empiricists, he did not see how we
could know anything substantial about what was out there, independent of
us, without letting it impinge on us through experience. He resolved the
apparent conflict here by arguing that the *a priori* knowledge in question
was not after all knowledge about what was out there, independent of us.

Kant's picture was as follows. When what was out there did impinge on us through experience, in such a way that we came to acquire empirical knowledge, this was only because we had certain epistemic faculties that made us appropriately receptive. Through these, we ourselves made a contribution to the shape of our experiences. We provided ourselves with a kind of framework through which to view things, with the result that whatever we experienced we experienced as having quite definite structural features. The *a priori* knowledge in question pertained to these features.

One aspect of this, according to Kant, was that we operated with certain concepts that were not simply read off from experience but rather were part of the framework. (The concept of the infinite was one. Others were such fundamental concepts as those of substance, causality, and number.) To this extent he was in line with the rationalists as against the empiricists. But, although he believed that we could use some of these concepts to think about things that utterly transcended our experience, he was in line with the empiricists in denying that we could use either them, or any other means, to gain substantial *knowledge* about what transcended our experience. Precisely their function, from an epistemic point of view, was to make experience possible. There was no prospect of putting them to epistemic use beyond this function.

Another, related aspect of his picture was that we could only ever know how things appeared to us, never as they were in themselves. We could never get 'behind' the framework. Kant accepted the kind of distinction between appearance and reality that we have seen so many times already in this enquiry. Moreover, he accepted it in as radical a form as anyone. For even space and time were part of the framework. Not even these were features of 'things in themselves'.

The bearing of this on the conflict between morality and science was as follows. Science was concerned with the natural world of space and time, the world of appearances. Within this world everything was subject to inviolable causal laws, and there was indeed no room for freedom. But we *were* free in ourselves. Of course, we could not properly understand this. But our innate moral sense made us aware of it, and aware also that, in some (to us incomprehensible) way, it was because of how we exercised our freedom that we appeared to ourselves, spatio-temporally, in the particular way that we did. It was only when we assumed that our spatio-temporal nature was all there was to us, in other words when we mistook appearance for reality, that this seemed incompatible with the dictates of science.

Kant held that our freedom was essentially bound up with our rationality. It consisted in our being able to put our rationality to practical use – being able to act in accordance with the laws of reason. The fact that we did not always do this, or even want to do this, was a result of our not being able to see things for what they really were. With our limited and conditioned view of things we sometimes found ourselves succumbing to the narrowly spatio-temporal appeal of irrational courses of action, even though they went against our true will. This was why, in Kant's view, the laws of reason presented themselves to us as moral obligations – not necessarily what we wanted to do but what we 'ought' to do.[3] It was when we did as we ought, and thus acted rationally, that we were being truly and completely free. Our actions were not then constrained, or befuddled, or answerable to any higher authority. They had their source in our very being, in something autonomous, self-sufficient, complete, and unconditioned. The importance of this, in the context of our enquiry, is that it points to what was, for Kant, a deep metaphysical infinitude in us – the metaphysical infinitude of reason.

But in order to understand the crucial role that Kant plays in our drama, we must start elsewhere.[4]

2 The metaphysically infinite and the mathematically infinite

Kant's greatest contribution towards an understanding of the infinite was a particular conception of how the metaphysically infinite and the mathematically infinite bore on each other, or so I shall argue. This had its roots in something that has already emerged: Kant's belief, human rationality notwithstanding, in our own metaphysical finitude.[5]

He believed that we were (metaphysically) finite beings in a (metaphysically) infinite world. We were part of a world that, in its own self-contained totality, was out there, independent of us, an absolute complete unified whole. This was precisely why we had to let it impinge on us in order to know anything substantial about it. We had to be receptive. (We differed in this respect from God, Who, in His infinitude, actually created what He knew through knowing it.)[6]

But Kant believed that we could never be receptive to the whole. Whatever we received was itself partial and finite, something particular impinging on us from without, something *conditioned*. He also believed that our receptivity must be self-conscious.[7] He thought we had to be able to *recognize* what we received as finite and conditioned, and this meant being able to receive something conditioning it (for example, being able to see what caused it). So whatever we came to know we had to be able to place in some suitable context, thereby coming to know more. But coming to know more in this way simply meant receiving more; and what we received would itself, along with what we had originally received, be a finite and conditioned part of the world. This set up an infinite regress: there was an endless series of conditions, each itself conditioned by some further condition in the series. But such infinitude was, of course, mathematical.

Kant's metaphysically infinite whole was thus presenting itself with a mathematically infinite aspect. (For Kant, the mathematically infinite was to be understood in much the same way as it had been understood by Aristotle – as the untraversable, or uncompletable.[8]) In this respect his system contained nothing new. Indeed it conformed to the recurring pattern which was mentioned in the last chapter and whose most recent exponent had been Leibniz. But it was here, for the first time, that a deep rationale for the pattern could be found.

So too it was in Kant, for the first time, that an intelligible connection between the metaphysically infinite and the mathematically infinite was at last established. They could now be seen as simply two conceptions of the same thing. As we saw in the previous section, Kant believed that there were certain *a priori* concepts in terms of which we could think about both reality and appearance. The concept of the infinite was one of them. And the connections that we have just seen traced out explain its Janiform nature. It was to be characterized in metaphysical terms insofar as it applied to how things really were; it was to be characterized in mathematical terms insofar as it applied to how they appeared.[9]

Postscript: This is an apt point at which to say some more about how Kant viewed the use of *a priori* concepts in thinking about reality. When they were being used in this way, they became what he termed 'Ideas of reason'. In a way this was a Kantian term of art, though he himself insisted that he was meaning 'Idea' in none other than the way in which Plato had meant it (see above, Chapter 1, §4). At first blush this is puzzling. Kant's Ideas and Plato's Ideas look like rather different animals, particularly in relation to knowledge. Both Kant and Plato held Ideas to pertain to reality rather than the world of appearances (the world of space and time), but for Kant this meant that they took us beyond the reach of knowledge, whereas for Plato it meant that they themselves were the (only) true objects of knowledge. In fact, however, the two of them meant something different by knowledge. Kant meant something whose paradigm was scientific understanding, Plato something whose paradigm had much more to do with a practical use of reason. So they were very close. It was of crucial importance for Kant that Ideas should have a practical use. (He even sometimes spoke in this connection in terms of practical knowledge.) We shall see something of this below.[10]

3 The infinitude of the world. The antinomies[11]

The fact that the metaphysically infinite had to be presented to us with a mathematically infinite aspect in the way described above meant that the framework through which we viewed things had to be itself in some measure mathematically infinite. For we needed to have enough 'room' to be able to receive the conditions of anything we could receive. And indeed Kant did hold space and time to be mathematically infinite, both by addition and by division. This meant that there was always scope for the conditions of anything in space or time to be located before it, around it, or within it. The fact that space and time were mathematically infinite was one of those substantial truths that we knew *a priori*. For it was a feature of the framework through which we viewed things. We simply presented the framework to ourselves in that way.[12]

This meant that Kant parted company with Aristotle as regards space. But he also, in a way, parted company with Aristotle as regards time. For, in a way, he held the infinitude of time to be present 'all at once'. At any point in time we could reflect *a priori* on its structure as a whole and thereby present its infinitude to ourselves. Still, Kant's position was like Aristotle's, and indeed like that of the empiricists, in that he denied that anything we ever came across *within* the framework was infinite. Whatever we were given in space or time was finite.

What about the physical world as a whole though? This was the question that had earlier threatened to embarrass Hobbes (see above, Chapter 5, §2), and now it threatened to embarrass Kant. For surely the physical world as a whole was something that we were given in space and time, and yet one of the main points of the argument sketched in the previous section seemed to be that, given any finite part of it, there must always be more to come (so it must be infinite).

Kant's riposte was the same as Hobbes'. He denied that there was any such thing as the physical world *as a whole*; there were only its parts. The point was this. Anything physical was ultimately mere appearance, so its existence depended on its being capable of being given to us in experience. The physical world as a whole was not capable of being given to us in experience. So there was no such thing.[13]

Of course, this ran counter to a deep and natural conviction, but Kant would have been the first to admit it. He had his own explanation for how the conviction arose. He held that we were subject to a natural and inevitable illusion, which was grounded in a conflation of appearance with reality. We knew that the conditioned only made sense given the unconditioned. For ultimately there had to be something complete and self-sufficient (something metaphysically infinite). So, given what was conditioned in space and time, and taking this to be what was ultimately real, we were disposed to think that there must also be something unconditioned in space and time, which we were capable of being given in the same way. But this was essentially to assume the existence of the physical world as an unconditioned whole.

In effect we were taking an Idea, our Idea of the unconditioned whole, and illegitimately trying to apply it within experience. Kant did however acknowledge that the Idea had a legitimate *practical* use pertaining to experience. (This bears out what was said in the postscript to the previous section.) For it was true that, given any finite chunk of physical reality, there was always more to come, namely a condition of it. So we could use the Idea to frame what Kant called a *regulative principle*: we could enjoin ourselves never to give up in the search for the conditions of things, by entertaining the Idea of the series of conditions laid out in advance as an unconditioned whole. Kant put it like this: we could, and indeed should, proceed, in scientific enquiry, *as if* the physical world existed as a whole.[14] The fact remained, on his view, that it did not.

This put him in a position to settle some of the long-standing debates that had been raging about the infinitude of the physical world, debates to which we have been witness. These were: the debate about whether the physical world was infinitely old; the debate about whether it was infinitely extended in space; and the debate about whether it was infinitely divisible. These arose only on the assumption that there *was* a physical world, existing as a whole (that is, capable of being given to us in experience as a whole). Granted this assumption, it was a perfectly legitimate question whether this world was infinite or finite in each of the specified respects; and it had to be one or the other. But then, on Kant's view, it was inevitable that irresoluble controversy should arise. There were reasons both why it could not be infinite (basically because we could not be given anything infinite in experience) and why it could not be finite (basically because we could not be given anything in experience that did not have further conditions that we were also capable of being given).[15] Rather than take up the cudgels for either side in these ill-conceived disputes, we should drop the common underlying assumption. In each of the specified respects the world – as a whole – was *neither* infinite *nor* finite. It did not exist.

Kant culled from the history of the topic what he took to be the central arguments in these vain controversies, and then laid them out alongside one another, so as to display the dialectic from his own impartial standpoint. Those concerning the world's infinitude by addition constituted what he called the first antinomy, those concerning its infinitude by division what he called the second antinomy.[16] Each argument contained a 'core' – a proof that the world could not be infinite, or finite, in one of the specified respects - which Kant took to be decisive. But each argument also contained an additional surreptitious move, based on the assumption that there *was* such a world: namely, that since it was not infinite, or finite, in the relevant respect, then it must be the other. And this is what Kant rejected. The 'cores', on this understanding, were not in conflict. In fact they could be combined, precisely to undermine the offending assumption. Kant was here playing his quintessential role as conciliator. He was putting paid to two thousand years of futile controversy.

Let us address the antinomies in turn, focusing on these 'cores'.

The Infinite

(i) The first antinomy

(a) *The proof that the world cannot be infinitely old, or infinitely big*
There are two halves to this proof, one for the temporal case and one for the spatial case.

(α) *The proof that the world cannot be infinitely old*: The infinite is that which is uncompletable, or untraversable. But the world's history, up to any given moment, has, by then, been completed, or traversed. So the world cannot be infinitely old. [Comment: This proof does not apply to time itself, because time is not, in the same way, traversed; we simply present it to ourselves as infinite.]

(β) *The proof that the world cannot be infinitely big*: Suppose that the world *were* infinitely big. Now it would have to be capable of being given in experience – either all at once, or 'through the completed synthesis of its parts.' The first of these would be impossible, given its infinite size. But the second, which would take infinitely long and thus require the world's being infinitely old, would likewise be impossible, given (α). So the world cannot, after all, be infinitely big.

Sources: Both halves of this proof were grounded in that recoil from the actual infinite which had dominated thought about infinity ever since the time of Aristotle. We saw it in Aristotle himself, in many of the medievals, and in the empiricists. Significantly, however, Kant was siding with those who opposed Aristotle and recognized a problem for infinite history as well as infinite bulk. Indeed, as we see, he took the temporal half of the proof to be more basic than, and to underlie, the spatial half.

Comment: In his own comment on this proof Kant rejected one popular account of the infinite which he might have been tempted to invoke here, an account which was part of the received wisdom of his day. (We saw it in Locke.) According to this account a magnitude is infinite when its measure, as a multiplicity, is incapable of increase. In Kant's view this made it too easy to rule out infinite magnitudes, because *any* multiplicity is capable of increase. (For example, however many days the world has existed, it will have existed one more tomorrow.[17]) He was happy to accept something that we saw earlier thinkers such as Philoponus and Locke balking at, namely that one infinite multiplicity could be greater than another. But, like them and so many others, he refused to license talk of infinite *number*, a number being what can be arrived at by a finite process of counting.[18] This was part and parcel of his own – Aristotelian – conception of the mathematically infinite, as the uncompletable or untraversable.

(b) *The proof that the world cannot be finitely old, or finitely big*

Time and space themselves are infinite. They are also homogeneous. It follows that there could be no reason for – indeed nothing could count as – a finite world's occupying any determinate position in either of them: there is no relation in which a finite world could stand to empty past time or to empty space which would *amount* to anything. The world is not, therefore, either finitely old or finitely big.

Sources: This proof was taken from a tradition which, as we saw, went back to Aristotle and had continued through to Bruno and then to Leibniz (though in Aristotle's case, only in support of the hypothetical: that *if* space were infinite – which Aristotle denied – then the world could not be finitely big). In aligning himself to this tradition, Kant was going beyond his own abstract principle that whatever is given in experience must have a condition that can also be given in experience. He was rehearsing a popular argument that bore specifically on time and space.

Comment: In the temporal halves of both (a) and (b) Kant explicitly confined his attention to the world's history rather than its future. This was because, in his view, a thing's temporal conditions necessarily preceded it. (There was no backward causation.) So assuming the existence of an unconditioned series of conditions only involved looking back to what grounded the present.[19] This ties in with the curious asymmetry that I alluded to at the end of Chapter 2. An infinitely old world strikes us as more problematical than a world with an infinite future, though it is very hard to say why. Of course, the asymmetry may simply be due to the fact that we share, with Aristotle and Kant, a sense of the infinite as the uncompletable or untraversable, though many would now regard these overly temporal characterizations (with their own past/future asymmetry) as question-begging. Perhaps, instead, we have a sense of time as cumulative, so that an infinite past, unlike an infinite future, would deliver an unacceptable actual infinite in the present.[20]

(ii) The second antinomy

(a) *The proof that what is composite cannot be divisible without limit*

Whatever is composite has parts; and some of these must survive a complete division or decomposition of it. (For there must be more to it than *just* its composition. There must also be that which it is ultimately composed *of*.) But if it were divisible without limit, then it could, at least in thought, be divided without limit, in which case no such parts would survive. Nothing composite, then, can be divisible without limit.

(b) *The proof that what is composite cannot have parts that are simple or indivisible* (*cannot be divisible only up to a limit*)

Whatever is composite is spatial. Likewise all its parts. But given that space itself is divisible without limit, each of those parts must itself have parts; for it must have parts corresponding to the parts of the space which it occupies. It follows that none of the parts – none of the parts of what is composite – can be simple or indivisible.

Sources: These proofs were more or less those that Aristotle had put forward for matter's being infinitely divisible in one sense but not in another. Versions of them were probably to be found in Zeno (in his case, fuelling Eleatic hostility towards the very idea of the composite – 'the many'). And they had recurred many times subsequently in thought about the infinite, for example in the medievals.[21]

General Source for Both Antinomies: Each of the antinomies had also, and most recently, taken the form of a debate between Newtonians and Leibnizians; and this was almost certainly what was firing Kant most – more than any of the sources cited above. The proofs labelled (a) in each antinomy had been urged by Newtonians, those labelled (b) by Leibnizians. The different positions concerning infinite divisibility had been set forth in the correspondence between Leibniz and Clarke that I mentioned in Chapter 5.[22] (There is a terminological problem here, due to a slight idiosyncrasy on Kant's part. Newtonians he described as 'dogmatists', Leibnizians as 'empiricists'. But the terminological discrepancy is not a complete surd. Part of the explanation for it pertains to how these two antinomies tied in with two others that he discussed, concerning freedom and God.[23])

These, then, were the proofs which Kant assembled from the history and which, on his conciliatory understanding, served to demonstrate the non-existence of the physical world as a whole. Indeed they also, on that same understanding, provided important indirect support for his philosophical system – in particular for his distinction between appearance and reality.[24] It is a significant and closely related fact that Kant, like many of his predecessors (though not, importantly, Aristotle), took empirical considerations to be irrelevant to these issues. Given that, for him, questions about the age and size of the world were rooted in an Idea of reason (albeit wrenched from its legitimate use), it followed that only *a priori* considerations could pertain to them.

Nowadays questions about the age and size of the world seem quite legitimate; and, unpersuaded (for the most part) by the proofs in the antinomies, we have come to recognize the important bearing of empirical evidence on them. It no longer appears to be the philosopher's prerogative to tell us (say) whether the world is infinitely big, or whether space and time themselves are for that matter. We feel that we can and should turn to the physicist instead (see below, Chapter 9, §2).[25]

It by no means follows that we have nothing to learn from Kant here. In Part Two I shall try to show how Kantian considerations can be brought to bear on fundamental *a priori* issues concerning the infinite that still exercise our thought. But there are lessons already to hand. Many of his views concerning the infinitude of the physical world had a very Aristotelian flavour and we can use them to sharpen our understanding of the distinction between the actual infinite and the potential infinite. Consider, for example, Kant's views concerning the infinite by division, as they arose from the second antinomy. He denied that, whenever we were given something physical, or composite – for Kant these amounted to the same thing – we were also thereby given the series of its ever smaller constituents as an unconditioned whole: but he conceded that division could never come to a halt. This was very much like denying that what was composite had an actual infinity of parts but conceding that it had a potential infinity of parts. Similarly with his views concerning the infinite by addition. Here it was very much as if he was saying that the physical world was potentially, but not actually, infinite: its infinitude could emerge in endless processes of enquiry but could never be finally grasped. It is important, however, that Kant was not exactly saying this. He still refused to accept that what was physical could exist unless it could be given in time. Being given *over* time was not good enough. So for Kant, the physical world did not exist even as a potentially infinite whole. (On the other hand, we must not forget the important concession that he made when he licensed a regulative use of our Idea of the infinite whole. I shall discuss this further in Part Two (see below, Chapter 14, §§2 and 4, and Chapter 15, §3).)

Kant's most important lesson for our understanding of the infinite, however, concerns the bearing of the metaphysically infinite on the mathematically infinite. What has emerged so far is the following fundamental idea. The metaphysically infinite world must appear to us with a mathematically infinite aspect; but not *as* mathematically infinite, because not, in its unconditioned entirety, as anything at all.[26]

4 The infinitude of reason

Something similar emerges in Kant's treatment of the metaphysical infinitude of reason, to which we now return.

For Kant this was again a feature of reality that we could not know as it was in itself. But unlike the metaphysical infinitude of the world, it was not something outside and independent of us. It lay deep within us. We had a direct, non-discursive awareness of it based on our moral sense, our sense of what was right and of our own freedom to practise it. Kant's most eloquent expression of this came in what is probably the best-known passage in his entire corpus, that which I have used as a frontispiece for this book.[27] In this passage he spoke revealingly about 'true infinity', drawing a contrast between that and the infinity of endless progression through space or time. It had been principally to safeguard our sense of this true infinity that Kant had developed his philosophical system in the first place, curbing the threatening pretensions of science. He wanted to give us scope to acknowledge, beyond the world of space and time, the metaphysical infinitude of our own rationality, and indeed the metaphysical infinitude of God, Whose existence, for Kant, was an object of faith that was inextricably interwoven with our (fully developed) moral sense. 'I have . . . found it necessary to deny *knowledge*,' he wrote, 'in order to make room for *faith*.'[28]

There were elements in this picture which, as we have seen, aligned Kant to the rationalists. He believed that we had a rational Idea of such infinitude, the infinitude of our own rationality, even though we could never meet it in experience. Indeed the grander and more magnificent what we did meet in experience, the more we were reminded of, and came to appreciate, what was infinitely more grand and more magnificent (what was 'absolutely great') in our Idea.[29] (Recall how Pascal had said that man, through his rationality and thinking, could achieve a nobility and dignity that transcended his finitude and thereby touch the infinite (see above, Chapter 5, §1). Kant would certainly have applauded that.)

None the less, Kant thought that we had to be able to see the infinitude of our rationality *in terms of* the world of space and time where its effects showed up.[30] Only then could we have any kind of grip on its consequences for us as finite beings. We needed to be able to recognize what behaviour of ours, in the here and now, would count as putting our rationality to practical use. This is why the laws of reason presented themselves to us as moral obligations to act in particular ways in particular circumstances.

But there was a problem here. These obligations were supposed to have an impact at particular *times*. How could they, though, given that rationality was itself fundamentally atemporal? Consider: all the wrong that we had done in the past surely showed that we had, in ourselves, for whatever reason, abnegated our rationality; there was nothing we could any longer do about it. So how could these obligations continue to *direct* us? Why did we not simply bow to them in abject contrition?[31]

A Kantian answer to this question would run, I think, as follows. Whenever the implications of our rationality were being spelt out for particular circumstances, what was being spelt out was how the infinite appeared from a particular point of view – not just a part of the infinite, as in the case of the world out there which we had to let impinge on us, in a piecemeal way, in order to have any sense of it. The very infinitude of the moral law had to be focused at each point from which it was viewed, irrespective of what was known about how things were, had been, or would be elsewhere. At any time, we had to feel the full force of morality, regardless of the wrong we had done in the past.

But this still seemed problematical. Such wrong seemed to point to an atemporal irrationality in us – an atemporal sinfulness – that meant that we were now past our own redemption. And this seemed to mean that the force of morality on us had now after all been mitigated. Kant had a solution to this problem which also involved him in his most fundamental departure from Christian orthodoxy. He held that we *could* atone for our own sin, or at least that we had to believe that we could. But this required time, time enough to live lives of which our past wrong-doing could become a negligible part and so in effect be obliterated. In other words it required infinite time. We had to believe that we were immortal. The metaphysically infinite was again presenting itself to us with a mathematically infinite aspect.

Not that we could see our future lives spread before us in their infinite entirety, or could be said to *know* that we were immortal. Kant was once again talking about a regulative principle. We had to act *as if* we had infinite afterlives.[32]

There was much here for later thinkers to take issue with and to disparage, as we shall see. But their disparagement could not mask the importance of Kant's thinking for an understanding of the infinite in its various different guises. Here, perhaps, was the most important attempt in the history of philosophy to engage with human finitude.

CHAPTER 7

Post-Kantian Metaphysics of the Infinite

I heap up monstrous numbers, pile millions upon millions, I put æon upon æon and world upon world, and when from that awful height reeling, I seek Thee, all the might of number increased a thousandfold is still not a fragment of Thee. *I remove them and Thou liest wholly before me.* (Albrecht von Haller)

Now, little ship, look out! Beside you is the ocean: to be sure, it does not always roar, and at times it lies spread out like silk and gold and reveries of graciousness. But hours will come when you realize that it is infinite and that there is nothing more awesome than infinity. (Friedrich Nietzsche)

1 Hegel

Kant's influence on subsequent philosophy was immense. Metaphysical thought about the infinite in the next two centuries was to be considerably shaped by his vision of man as a radically finite creature cast into an infinite world and yet exalted by the infinitude of his own freedom. The influence is especially clear when we turn to one of the greatest of his successors, the German philosopher G. W. F. Hegel (1770–1831).

In certain respects Hegel played Aristotle to Kant's Plato. He was deeply influenced by Kant's philosophical system, but there was much that he sought to repudiate. In particular he wanted to challenge Kant's distinction between appearance and reality. The idea of 'things in themselves' beyond our epistemic grasp was an anathema to Hegel. Reality, for Hegel, was not something *underlying* appearance, but something essentially manifest *in* appearance – in the world of space and time. (Here, as elsewhere, his thinking was very reminiscent of Spinoza's.) Kant had been right to see reality as a self-contained and absolute whole. He had also been right to see rationality and freedom in these terms. But he had been wrong to try to set this apart from what we can encounter and come to know about in experience. It was in this light that Hegel attempted to reach a new and deeper understanding of the infinite.

96

Aristotle had been in a similar position. He too had rejected the reality/appearance distinction of his predecessors, including Plato, and had tried to reassess the infinite accordingly. But whereas Aristotle thought that what was required was a new conception of the infinite, as something that could be understood in spatio-temporal terms, Hegel thought that what was required was a new conception of the spatio-temporal, as something that could be understood in terms of what was now recognizable as the truly infinite. For he felt that Kant's (essentially metaphysical) conception of the truly infinite was fundamentally right. (His understanding of Kant was broadly the same as that which I tried to present above.)[1]

Hegel was with Kant in recognizing the truly infinite in the absolute and complete unity of the whole. He was also with Kant in recognizing it in the free, autonomous self-sufficiency of reason. But what he did was to run these together in a way that Kant had not. He believed that what was real *was* ultimately what was rational. Reason was the ground of everything. Whatever happened could be understood as the activity of a kind of world-spirit, and this spirit *was* self-conscious reason. The infinitude of the whole and the infinitude of reason were of a piece.

A corollary of this was the essential unity of everything: nothing ultimately made sense in isolation from anything else. Truth itself ultimately resided in the whole. It too was metaphysically infinite. Remember how Nicholas of Cusa had claimed that only what was itself the truth – the infinite – could know the truth; anything else, that is anything finite, could only approximate to a knowledge of the truth (see above, Chapter 3, §4). Hegel accepted something very like this. This deepened the connection between the infinite and self-sufficiency. The truly infinite was both the knower and the known. There were also connections with freedom. It was precisely through such self-knowledge that the world-spirit was free to act in accordance with its own rational principles. The metaphysical infinitude of reason that Kant had highlighted was at the heart of Hegel's system.

What then was it for something to be finite?

It was for it to be a mere aspect of the whole, something limited and up against an 'other'. A finite thing's 'other' both defined and negated it. It determined both what it was, and what it was not. The thing itself was finite precisely because it could be delineated and set apart in this way. This was why Hegel (just like Spinoza (see above, Chapter 5, §1)) held that nothing could exist that was not part of, or an aspect of, the infinite. If the infinite did not embrace the whole of reality, then it too would have some 'other' that would serve to define it, and negate it, and thereby make it finite. In particular the infinite was not to be *contrasted* with the finite. It incorporated the finite. It held together opposing strains of finitude in the unity of the whole, that is in itself. This was one reason why Hegel could not accept a radical appearance/reality distinction. The real – the infinite – could not be seen as something set apart from appearance, or to be contrasted with appearance, for then it would be finite.

The German philosopher and theologian Friedrich Jacobi (1743–1819) had taken up the idea that, because we ourselves were finite and therefore operated with finite concepts and finite categories, our discursive thinking inevitably perverted and misrepresented the infinite (that is, God); our awareness of the infinite had to be direct and intuitive.[2] Some of Jacobi's thinking was endorsed by Hegel. But there were aspects of it that he could not accept.[3] He could not accept that our attempts at discursive understanding were altogether isolated from direct insight into the infinite. On the contrary, they were stages in the development of the infinite's own self-consciousness. True, our own knowledge was always partial, conditioned, and incomplete. Whatever we grasped, there was always more to grasp. But we always *could* grasp more. The process could continue, as part of the very process of the infinite's coming to know itself. Thus it was that the infinite had an inner articulation that could unfold for us as we came to know more and more about it. Hegel, in a very similar way to Kant, believed that the infinite whole presented itself to us with a *mathematically* infinite aspect.

We have yet to look into Hegel's attitude to the mathematically infinite. But first we must broach the question of how, in the light of this, he regarded the physical world (as a whole). Did it follow that *that* was mathematically infinite? What did Hegel make of Kant's antinomies? (In answering these questions, we shall see something of his rather idiosyncratic attitude to the spatio-temporal.)

The first thing that needs to be said is that he found the proofs in the antinomies seriously question-begging, at least as presented by Kant. But that did not really matter. For he agreed that they pointed to what were genuine reasons for regarding the whole physical world both as finite and as infinite. He could not avail himself of Kant's own solution to this problem, for two reasons. First, he did not share Kant's belief that what was physical was ultimately unreal. (This had enabled Kant to deny that the physical world existed as a whole.) Secondly, he could not accept that, by locating the contradiction in (a mistake of) reason rather than in reality, Kant had made any kind of advance. For Hegel, reason and reality were one. Whatever problem there was with the physical world's being both finite and infinite was, on his view, every bit as much a problem with our being led to believe, by pure ratiocination, both that it was finite and that it was infinite.[4] He needed some other way of addressing the antinomies.

He chose a way that was breathtakingly cavalier, more so even than that of Hobbes and Kant when they had severally denied that there was any such thing as the physical world as a whole. He accepted the contradictions. This was reminiscent of what Nicholas of Cusa had done much earlier.

He had argued that the infinite was all it could possibly be and that it need not be free of contradiction; for it was beyond our finite categories. For Hegel, the infinite embodied a co-existence of supplementary but opposed finite aspects. Whenever we tried to think about it in finite terms, as in the proofs in the antinomies, these oppositions were liable to come to light. They were oppositions that, in Hegel's developed system, were 'worked out' in time – rather as *to apeiron* of Anaximander had given rise to a series of conflicting opposites that were 'worked out' in time and destined, ultimately, to be resolved in the unity of the whole (see above, Chapter 1, §1). The antinomies pointed to contradictions inherent in how the world was presented to us.[5]

These contradictions had their seat in the mathematically infinite. We must now turn to Hegel's view of this. He held that the concept of the mathematically infinite was rooted in our understanding. It arose as soon as we started trying to think discursively about the infinite, partly indeed because of how the infinite was presented to us. But the mathematically infinite was what Hegel described as a spurious, or bad, infinity – a mere succession of finite elements, each bounded by the next, but never complete and never properly held together in unity. It seemed at turns nightmarish, then bizarre, then simply tedious, but always a pale, inadequate reflection of the truly infinite. Moreover, it was a kind of infinity that *could* be set in contradistinction to the finite (unlike the truly infinite). And whenever it was, it too seemed to be finite, because it was just one aspect of the truly infinite whole; but then again infinite, because it was supposed to take us beyond the merely finite; but then again finite; and so on *ad infinitum*. This oscillation revealed both the intrinsically contradictory and unstable nature of the spuriously infinite, which had shown up in the antinomies, and, of course, the fact that the concept had unfurled before our understanding in a spuriously infinite way.

The upshot of all of this, we now see, was one of the great ironies in the history of thought about infinity. By dint of a similar metaphysical recoil, Aristotle and Hegel arrived at positions that were the exact mirror-image of each other. Aristotle had explicity repudiated a metaphysical conception of the infinite in favour of a mathematical one. Hegel explicitly repudiated a mathematical conception of the infinite in favour of a metaphysical one.[6] They represent polar opposites in this dialectical history.

Aristotle, complaining that the infinite was the exact opposite of what people said it was, had gone on to develop a conception of the infinite that was destined to dominate subsequent thought on the topic. It was not, he said, 'what has no part outside it . . . but what always has some part outside it,' in other words a never-ending progression in which 'one thing is taken after another, what is taken being always finite, but ever other and other.'[7] For Hegel, this was completely back to front: the truly infinite was *precisely* 'what has no part outside it'. He wanted to return to the diametrically opposed view. The correct geometrical image of infinity, for him, was not an endless straight line but a circle.[8] (This was reflected, interestingly, in his philosophical method. The image of a circle signalled a particular kind of understanding that he strove for: that which resulted from one's following through the implications of something until one had been brought back to one's starting point and could make sense of it in the entirety of its context. Such understanding found its fullest expression in that complete, rational insight into the infinite whole which the infinite alone could achieve.)

The real problem with the mathematically infinite, for Hegel, was that it never really got beyond the finite. It was an *attempt* to reach the genuinely infinite, just as Kant's argument for immortality had arisen from an *attempt* to recapture the infinitude of reason. But all that resulted was 'a never-ending approximation to the law of Reason.'[9] This left us trapped with a perpetual 'ought'. For there was a constant tension in this endless progression between where we aspired to be and where we actually were. And this helped to make graphic the way in which the infinitude of reason presented itself to finite beings such as us as an unrelenting obligation whose force never diminished. Hegel believed that the human will was itself finite, defined by its 'other', nature. This made it unable to achieve its own ends directly, always feeling that there was more to be done. The mathematical conception of the infinite provided a kind of image of this. But despite that, and despite the powerful grip that the conception had retained ever since the time of Aristotle, it was not, for Hegel, in any sense a true conception of infinity.[10]

In the nineteenth and twentieth centuries there were various currents of metaphysical thought about the infinite that were more or less directly inspired by Hegel or had Hegelian overtones. At the same time Kant's (and Hegel's) conception of human finitude provided a continuing focus, something that will be especially clear in the work of the existentialists. It is to this and other post-Hegelian thought that we now turn.

2 Currents of thought in post-Hegelian metaphysics of the infinite I: the 'metaphysically big'

The Scottish philosopher Sir William Hamilton (1788–1856) took up a point that reinforced what Jacobi and Hegel had said about our inability to think discursively about the infinite. To think about anything, Hamilton said, was to think about it as a thing of a certain *sort*. This already involved classification and the imposition of conditions. We could not think properly about the unconditioned, or the infinite. Or at least, we could not think *what* it was, though we could think *that* it was.[11] (Here, of course, he was skirting one of the basic paradoxes of thought about the infinite.)

Karl Solger (1780–1819), a German romantic philosopher, had the related belief that man, in his finitude, could only ever grasp fragments of reality. But man had a desire for the infinite, and, just as God had sacrificed Himself in creating the finite, so man must make a sacrifice in returning to the infinite. Solger believed that philosophers, artists, and moralists were all alike concerned with the reconciliation of the infinite with the finite.[12]

The Italian-born philosopher Baron Friedrich von Hügel (1852–1925), who spent most of his life in England, was another who took up these themes. He agreed that we longed to transcend our finite limitations, even though this was ultimately impossible. All our achievements were limited and conditioned by ignorance and finite modes of thinking. Yet we had an awareness of the infinite (or of God). This was precisely why we were dissatisfied with the finite. We found that the infinite impinged on our experiences, our emotions, and our wills, and acted as a source of inspiration. Indeed there was a sense in which we could grasp it, though this was as much a matter of the will as of cognition. On von Hügel's conception, the infinite had to be something spiritual – but something that could be present in finite spirits such as ours.[13]

When we turn to the English philosopher F. H. Bradley (1846–1924), we see Hegel's influence very clearly. Bradley had a metaphysics that was in many respects pure Hegel. There was also something very Eleatic and Neoplatonic about it. He believed in an Absolute (a term that Hegel had also used), a complete self-contained unity, which appeared to us as spatio-temporal. But its appearances were riddled with contradictions. There were two main reasons for this. First, they constituted a plurality. (The One appeared to us as a many.) But sense could ultimately be made only of unity. And secondly, they presented themselves as mathematically infinite. (Yet again we see this recurring pattern.) But Bradley, like so many before him, recoiled from the mathematically infinite, or at least from the mathematically infinite when presented as actual, believing that no sense could be made of it. As we contemplated the full extent of the world in space and time, we were thus led to believe – as Kant, Hegel, and indeed Zeno had all insisted – both that it was infinite and that it could not possibly be infinite. Such contradictions were ultimately reconciled in the unity of the Absolute, though in a way that we could not understand. We could think only in terms of 'the other', in its infinite richness, going beyond our finite categories – beyond our 'finite centres of experience'.[14]

These ideas were later expressly challenged by Josiah Royce (1855–1916), an American writing in a similar tradition but with much less antipathy towards the actual (mathematical) infinite. He asked us to consider why the actual infinite had continually been such an object of abhorrence. The paradoxes of the infinitely big seemed to be the real bugbears. But technical work by Dedekind, Cantor, and others (which we shall be looking at in the next chapter) had finally shown, in Royce's view, how these paradoxes could be avoided. There was nothing incoherent in the idea that a part of something should be perfectly correlated with the whole. He had his own elegant illustration of this. He argued as follows.

Royce's argument: A perfect map of some region might itself form part of the region (for example, it might be set out on some flat piece of ground). Then it would be a part of the region that was perfectly correlated with the whole.

So long as we stopped trying to think about the actual infinite in finite terms, then it (the actual infinite) made perfectly good sense, and we could (and did) accept the existence of infinite wholes. True, any whole had to be determinate. But it did not follow that it had to be limited. The natural numbers, for example, together constitued a quite determinate totality. Similarly with reality itself, the Absolute, which in Royce's view was both a one and a many. We could never fully grasp it, admittedly, but we could get nearer and nearer to a grasp of it – the kind of thing that philosophers had been urging ever since the time of Nicholas of Cusa.[15]

A. E. Taylor (1869–1945), an Englishman in the same tradition, heavily influenced by both Hegel and Bradley, agreed with Royce that recent technical work had vindicated the actual infinite. But he did not think that this was enough to rescue the spatio-temporal world from contradiction. For he had an argument to show that its infinitude was of a kind to force a vicious infinite regress (roughly, regions of space and time were relations between regions of space and time). It must, therefore, be recognized as mere appearance.[16]

Because of his acceptance of the actual infinite, Taylor also felt able, interestingly, to endorse a Hegelian conception of true infinity. For if the truly infinite was that which has 'an internal structure which is the harmonious and complete expression of a single self-consistent principle,'[17] and if mathematical sequences such as

$$\langle 0, 1, 2, \ldots \rangle$$

or

$$\langle 1, \tfrac{1}{2}, \tfrac{1}{4}, \ldots \rangle$$

were viewed as definite and completed wholes, then they too could count as infinite, not because they were endless, but because they had an inner structure determining the derivation of each term from its predecessor. (The sequence

$$\langle 0, 1, 2, 3 \rangle$$

by contrast, stopped at a point that was quite arbitrary.) Thus friendliness towards the actual (mathematical) infinite had, unsurprisingly, undermined the metaphysical/mathematical distinction; unsurprisingly, because the actual (mathematical) infinite is essentially an amalgam.

Taylor's main focus, like Royce's, was the infinitely big. All the thinkers that we have looked at in this section were concerned in some sense with grandeur, what we might call the 'metaphysically big'. (I use this expression loosely. I do not mean to register any special connection with the metaphysically infinite, beyond being of metaphysical concern.) But the 'metaphysically small' also continued to arouse interest. Issues of continuity, divisibility, and simplicity were rife. We turn next to these.

3 Currents of thought in post-Hegelian metaphysics of the infinite II: the 'metaphysically small'

One of the reasons that continuity poses such a problem is that what is continuous (a line, say) seems to be both essentially a unity and essentially composed of parts. The German philosopher Johann Herbart (1776–1841), who exercised an influence on Bradley, took up this problem and argued that, because the continuity we confront in experience did indeed involve both unity and separation, in a contradictory way, this was one reason for holding space and time to be ultimately unreal.[18]

Much later his compatriot Edmund Husserl (1859–1938), one of the greatest philosophers of the recent past, took up the same issue. He said that the continuum of appearances had a unity that was for us unthinkable – not only with respect to its continuity but also with respect to its extent as a whole. But we could gain insight into this unity by means of a Kantian Idea of the infinite, not, indeed, by illegitimately applying the Idea within experience, but by exercising it to gain that special kind of insight that Ideas alone could afford. It was only in this way, Husserl argued, that we could so much as deny that the infinite continuum could be given with complete determinacy to a finite consciousness in the first place.[19] (Once again the paradoxes of thought about the infinite were surfacing.)

An almost exact contemporary of Husserl was the Frenchman Henri Bergson (1859–1941), for whom continuity was a major preoccupation. He held that mathematics involved a fundamental falsification of (both experiential and physical) continuity, because of its commitment to the notion of a point. It encouraged the idea that continuity could be built up out of points, whereas the most that could be built up out of points was a series of infinitely repeated discontinuities. Continuity was something basic, and points were just a mathematical fiction wrought from it. Moreover, temporal continuity was subject to a double falsification. For the notion of an instant compounded these difficulties by presupposing, or insofar as it did presuppose, a spatial conception of time. On Bergson's view, changes and movements, for example Achilles' run from *A* to *B*, were unanalyzable seamless wholes, simple and indivisible. When Achilles ran straight from *A* to *B*, it was true that he *could* have paused half way and thereby broken his journey. But in fact he did not. He ran straight there. Zeno's paradoxes of motion, and the other related paradoxes of the infinitely small, arose only on the mistaken assumption that such changes and movements could be broken down into parts (limitlessly) in the same way that spaces could.

In fact, Bergson believed that Zeno had performed a great service by drawing attention to an incoherence in the standard scientific conception of change, which was for him (Bergson) radically inauthentic and not the same as our pre-reflective conception. Pre-reflectively we had a direct, non-discursive, intuitive access to change: we just saw it. We saw it, moreover, as a fundamental feature of reality, as against what Parmenides, Zeno, and the other Eleatics had argued (and what had been argued many times since). Bergson also believed, incidentally, that we had the same kind of direct, non-discursive, intuitive access to what he called 'the absolute' or 'the infinite', and to our selves. Such access was to be contrasted with the kind of understanding which resulted from analysis and decomposition. It was when these were applied to change, and only then, that change appeared to be both a unity and a multiplicity, giving rise to irresoluble tension. There was no such tension when, by simple intuition, we immersed ourselves in 'the concrete flowing of duration' and observed the change for what it was.[20]

This idea that there is something primitive and unanalyzable about change, or more generally about continuity, has a long and distinguished history. There was something of it in Aristotle; and the brilliant logician Gödel, whose work we shall be examining in some depth later, distinguished between the set of points on a line and our intuitive concept of the line itself when he said:

> According to this intuitive concept, summing up all the points, we still do not get the line; rather the points form some kind of scaffold on the line.[21]

It is worth mentioning finally in this section how certain abstract versions of Kant's second antinomy lived on after him. The English philosopher J. M. E. McTaggart (1866–1925) said that it was a self-evident and ultimate principle that there could be no simple objects, either of perception or of any other kind: being an object (or a substance) necessarily involved having internal structure.[22] This was at roughly the same time as Wittgenstein was arguing that we could not so much as think about a world that did not contain simple objects.[23]

4 Currents of thought in post-Hegelian metaphysics of the infinite III: the existentialists

The philosophers traditionally classified as existentialists are united by a family of more or less directly related concerns and interests rather than by a settled body of doctrine. But one concern which was of central importance for all of them and which showed the continuing influence of Kant was human finitude.

It is the Danish philosopher Søren Kierkegaard (1813–1855) who is usually reckoned to be the first of the existentialists. For Kierkegaard human finitude was something basic. It was grounded in the fact that our understanding of things was inescapably limited and partial. But he thought that, in our relationship to God and in our awareness of infinite possibilities, we also had a sense of the infinite; indeed, we were ourselves in some measure infinite. These two conflicting elements in our being, if not properly synthesized, inevitably led to despair, as each abnegated the other. If they *were* properly synthesized, on the other hand, then we 'became ourselves'. And we were free to bring about that synthesis, by an ultimate, and ultimately criterionless, choice: to accept reconciliation with God.[24] (This was very similar to what we saw being argued a generation or so earlier by Solger, and to the ideas that were later taken up by von Hügel (see above, §2).)

This pattern of conflicting elements of finitude and infinitude in our being came to play a crucial role in later existentialist thought. It dominated the work of the German Karl Jaspers (1883–1969). Jaspers, who was one of the first philosophers to use the term 'existentialism', was heavily influenced by Kierkegaard. He was, like Kierkegaard, deeply religious. And, like Kierkegaard, he took human finitude to be basic. He believed that man was defined by his finitude and, in a very real way, aware of it: he was aware that he was dependent on his environment, that he was dependent on his fellow men, that he was limited in what he could know, and, most vividly of all, that he would one day die. But again, Jaspers recognized a dimension of infinitude within man. This was grounded in the fact that man knew himself to be free. There were Kantian echoes here of course. But Jaspers was alluding to something rather different from Kant. He was alluding to the infinite possibilities within man's finite boundaries to which his freedom gave rise. However, there was a sense in which this merely served to emphasize man's finitude. For such freedom had to be seen as a gift from without, not something that man, *in* his finitude, could possibly have conferred upon himself. Furthermore, whenever he freely adopted particular courses of action, some of the infinite possibilities had to be closed off. His boundaries could be seen closing in on them.[25] (Compare this with the Pythagorean and Platonic idea of the imposition of the *peras* on *to apeiron*.)

105

Man was radically finite then. But Jaspers believed that, in his very consciousness of this finitude, man was at the same time conscious of a contrasting and transcendent infinitude that lay beyond his horizon. This was something that he was for ever striving to gain. Through his freedom, and through faith, he could indefinitely tend towards it. But, precisely because he was ultimately finite, it remained for ever beyond his reach.[26] (Again we are reminded of Solger and von Hügel.)

Other existentialists adopted a much less religious tone and focused on human finitude much more for what it was in itself, as opposed to what it pointed to beyond itself. They portrayed human existence very starkly, in its own terms. Man was cast adrift in a world that was not of his own making. He was isolated and bemused. But he was also free, so far as his finitude allowed him to be, and so far as it gave such freedom any point. Thus, in their different ways, different existentialists were led to something that we have just seen in Jaspers: the idea that human beings were endowed with a dimension of infinitude, which, paradoxically, served to accentuate how far they were finite. Here once again is the tension that we saw in Kierkegaard. And it was understood by later existentialists in a very similar way. They related it to what they called the absurdity of human existence. Feelings of dread, guilt, despair, and anguish had their roots in these conflicting strands in our being, in the tensions between free choice and pointlessness, between potentiality and actuality, between expansion and constriction, between sense and nonsense, between the infinite and the finite.

The German Martin Heidegger (1889–1976), one of the great twentieth-century philosophers, is a key figure here. Heidegger was preoccupied with the fundamental nature of being. In this preoccupation he saw himself as going right back to the concerns of the early Greeks, indeed back to the question which, through Anaximander, initiated our own enquiry: what is the 'principle' of all things (the basic stuff of which all things are made)?[27] (Anaximander's own response to the question fascinated and captivated Heidegger. He wrote at length about the fragment which I cited in Chapter 1, §1.[28]) At the same time he was self-consciously and deeply influenced by German philosophy from the time of Kant. It was inevitable, then, that he too should become profoundly concerned with human finitude. Indeed it was Heidegger, more than anyone else, who was responsible for drawing attention to, and emphasizing, the importance of human finitude in Kant.[29]

His own portrayal of human finitude had many of the elements described above. He picked up on both the metaphysical and the mathematical aspects of it. The influence of Kant led him to probe the former.[30] But it also inspired his special interest in time, and this in turn induced a distinctive emphasis on the latter. For, like Kierkegaard and Jaspers, he took human mortality to be one of the chief marks of human finitude. Our time within the world was – mathematically – finite. For Heidegger this was

106

of an importance beyond exaggeration. But it was also to be understood in a radically new way.

It was important, because it provided a parameter for all our concerns, cares, projects, and aspirations, indeed whatever was of significance to us. It grounded many of those deepest conflicts in our existence described above. It gave our sense of fate and guilt all their poignancy (witness Kant's attempts, among many others, to wriggle free of it). Death, paradoxically, gave life its meaning.[31]

But it was to be understood in a new way, because it was not just that, for each of us, there would come a time, in the infinite expanse of time, when he or she would be no more. Rather, time itself, at the most fundamental level, what Heidegger called 'authentic' time, was finite. That is, each of us was given time *as* finite. This was the time defining his or her own existence. The time of ordinary, common-sense understanding, the time that was supposed to carry on endlessly after each of our deaths, was somehow abstracted from this (and it was one of the most urgent of philosophical problems to understand how). It was the basic finitude of authentic time, Heidegger argued, combined with certain fundamental asymmetries that he recognized between its past and its future, that imposed on our intuitive conception of time a sense of its 'passing' rather than 'arising'.[32]

Heidegger thus seemed to part company in a crucial respect with each of Bergson and Kant: with Bergson, in saying that it was our common-sense understanding of time that was radically inauthentic; and with Kant, in saying that time was ultimately given to us as finite. (But the contrast with Bergson may not be real. There is the question of how far what Heidegger called our common-sense understanding of time had already been tainted by scientific reflection.)

There was a very different emphasis in the writings of the great French novelist, playwright, and philosopher Jean-Paul Sartre (1905–1980). He urged that we would be finite even if we were immortal. For he understood human finitude more as Jaspers did. It was rooted above all in two things: in the fact that whenever we exercised our freedom, we at the same time excluded certain possibilities; and (relatedly) in the sheer brute contingency of our existence.[33] Furthermore, it was because of our finitude, Sartre argued, that things appeared to us as inexhaustibly rich. Their very objectivity meant that they could be viewed from infinitely many different points of view, not all of which we, in our finitude, could occupy.[34] This provided another glimpse, somewhat different from Kant's, into how the metaphysically infinite and the mathematically infinite came together.

The Infinite

5 Nietzsche

Finally in this chapter Friedrich Nietzsche (1844–1900), the brilliant German philosopher, himself sometimes classified as an existentialist, makes his own distinctive contribution to our drama.

Nietzsche reacted violently against much of Kant. He also reacted violently against Christianity. And the moral categories that were integral to both – sin, guilt, redemption, and the like (albeit handled in a highly unorthodox way by Kant) – were subject to some of his most challenging and withering iconoclasm. True redemption, for Nietzsche, was born not of guilt-ridden penitence or contrition. It demanded a triumphant affirmation of one's past. He described 'It was' as 'the will's teeth-gnashing and most lonely affliction.' But when the 'It was' was transformed into a 'Thus I willed it' – that was when the past was truly redeemed.[35] But the transformation was only possible given a creative act of the will, and this must be revealed in the continuing saga of one's life. One had to redefine and reinterpret one's own past *by* living a life in which one could come to accept it, and make sense of it, as an integral part. One's biography, or autobiography, became a work of fiction continuously in the writing. And so where Kant had talked in terms of overcoming one's past conduct, Nietzsche talked in terms of repossessing it. Where Kant had urged us to act as if our lives were part of an eternal progression to perfection, pictured as an infinite straight line that moved away from our past conduct, Nietzsche urged us to act as if our lives were part of an eternal recurrence, pictured as a circle that came back to it.

This doctrine of an eternal recurrence – the doctrine that every event in the universe will occur again in exact detail infinitely many times, just as it has already occurred in exact detail infinitely many times – has a long history. Some of the Pythagoreans believed it, and versions of it were held by later Greek thinkers, for example Empedocles.[36] Nietzsche himself may have believed it, in the sense that he may have thought that events literally recurred in this way. He certainly toyed with various arguments concerning the play of finite resources in infinite time, arguments whose upshot seemed to be that all of the finitely many things that could happen, and did happen, must eventually be repeated.[37] But of far more concern to Nietzsche was the impact of belief in the doctrine. What would be involved in living one's life *as if* it were true? One would strive to live in such a way that one could bear the infinite repetition. One would strive to affirm one's past, in just the way that Nietzsche believed one must. And so the doctrine of eternal recurrence enabled him to present his anti-Kantian, anti-Christian invective in a particularly graphic way. The joy or despair that the prospect of eternal recurrence afforded helped to give shape to the project of creatively making sense of one's life – past, present, and future. The sense was that much deeper, and that much more awesome, for

108

enjoying a certain infinitude. But as the image of the circle helps to show, this was beautifully ambiguous as between a metaphysical and a mathematical infinitude. The eternal recurrence had elements of both. Thus the topic of infinity proved to be (yet) another of the great philosophical perennials to which Nietzsche added his own curious and distinctive twist.[38]

CHAPTER 8

The Mathematics of the Infinite, and the Impact of Cantor

[If] a man had a positive *idea* of infinite ..., he could add two infinites together: nay, make one infinite infinitely bigger than another, absurdities too gross to be confuted. (John Locke)

For over two thousand years the human intellect was baffled by the problem [of infinity] ....

A long line of philosophers, from Zeno to M. Bergson, have based much of their metaphysics upon the supposed impossibility of infinite collections .... The definitive solution of the difficulties is due ... to Georg Cantor. (Bertrand Russell)

Apart from an anti-Aristotelian backlash among the medievals, spear-headed by Gregory of Rimini and partly followed through by the rationalists (see above, Chapters 3 and 5), the time up until the early-mid nineteenth century saw nothing but hostility towards the actual mathematical infinite. Some of the hostility was towards the mathematical infinite *per se*. We saw something of this in Hegel. But some of it was emphatically not that. It was hostility specifically to the *actual* mathematical infinite. Here, of course, the key figure was Aristotle, for whom the infinite was certainly to be understood in mathematical terms; what had to be resisted was the idea that it could be given 'all at once'. For over two thousand years this was the prevailing view.

As this view developed and achieved the status almost of orthodoxy, the notion of being given 'all at once' came to be understood in increasingly metaphorical terms. Often it just meant being a legitimate object of mathematical study in its own right. And, from that point of view, mathematics itself bore witness to the prevailing hostility. For the infinite never really was regarded as a legitimate object of mathematical study in its own right. True, it continually impinged on mathematical consciousness. But this was only because mathematicians, in their study of finite objects such as natural numbers and lines, would constantly step back and reflect self-consciously on the framework within which their study was taking place. Time and again the idea that there should be, say, infinite *numbers*, or any other precise mathematical tools for measuring and comparing infinite magnitudes, was subjected to ridicule.[1]

As regards the infinitely small, this was largely vindicated by the gradual refinement of the calculus, though Robinson's work on infinitesimals went some way towards challenging the orthodoxy (see above, Chapter 4, §2). Our main concern in this chapter, however, is the infinitely big. Here the main problem was undoubtedly with the paradoxes (the paradoxes of the infinitely big, that is). It was generally felt that, because of them, any attempt to talk about infinite magnitudes as completed wholes with determinate sizes was bound to lapse into contradiction and incoherence.

But why?

Consider the paradox of the even numbers. In general terms this shows something that was only to be expected, namely that we cannot straight-forwardly deal with the infinite as if it were just like the finite. More specifically it shows, or helps to show, that there are two families of criteria for comparing sets in size, which always co-incide when applied to finite sets but not when applied to infinite sets. There are the criteria that turn on whether the members of one set can be paired off with the members of another; and there are the criteria that turn on whether the members of one set all belong to another. Let us call these respectively 'correlation' and 'subset' criteria. Given that the infinite and the finite are *not* the same, and given that these criteria are easy enough to distinguish, what is the objection to our following Gregory's lead by simply describing the situation – quite consistently – in terms of this distinction? Thus we can say that there are as many even natural numbers as natural numbers in the 'correlation' sense, but there are fewer of them in the 'subset' sense. Provided that we always spell things out carefully in this way, there seems to be no threat of contradiction.

Of course, we might want to say that one of these senses is the 'true' sense, though explaining just what this means is liable to raise thorny issues in the philosophy of language. A much less ambitious move would be simply to *opt* for one of the senses and to stipulate that, when discussing the infinite, we are going to use the relevant phrases only in that way.[2] But whatever we do, so long as we make ourselves suitably clear, it looks as if we can address the paradoxes of the infinitely big without fear of contradiction. At the very worst, some of our intuitions will be under strain, but again that is only to be expected. And now it is not obvious why a rigorous, systematic, unified, mathematical theory of the infinite – the actual mathematical infinite – should not be available. If it is, that is if we can talk about the infinite in a way that is both coherent and precise, then we shall have the most effective possible vindication of the actual mathe-matical infinite against the weight of sceptical tradition.

111

The Infinite

This is the cue for Cantor. Georg Cantor (1845–1918), born in St Petersburg of Danish parentage – though he spent most of his life in Germany – was a mathematician of genius who devised a theory of precisely this kind. Only Aristotle and Kant are as important in the history of thought about infinity. In terms of his mathematical contribution to the topic, nobody can hold a candle to him. He was years ahead of his time. He single-handedly paved the way for a whole new branch of mathematical enquiry and research. I shall present some of his most important results in this chapter, though many of the main features of his work will be reserved for Part Two.

First, however, we must set the scene by looking at related work that was being carried out around the same time, both informing Cantor's work and heralding later fusions of ideas. Some of his arguments had already been put forward by Bernard Bolzano (1781–1848), a philosopher, theologian, and mathematician, who was born in Prague but spent much of his life in southern Bohemia. It is to his work that we turn first.

1 Bolzano[3]

Nineteenth-century mathematics was witness to the increasing importance of the concept of a set. (In fact this ground-breaking theory of Cantor's to which I have just referred, as well as being a theory of the infinite, was the first systematic mathematical theory of sets. I quoted Cantor's two famous definitions of a set in the Introduction, §3.) Bolzano was one of the first thinkers to promote the idea, later of such importance to Cantor, that application of the concept of infinity was first and foremost to sets. To describe something as infinite in a given respect was always to say, if only indirectly, that some particular set had infinitely many members. For example, to say that God was infinite in respect of knowledge was to say that the set of truths known by Him had infinitely many members.[4] And, *pace* Hegel, this meant that the mathematical conception of the infinite was precisely the right conception. For what did it mean to say that a set had infinitely many members? It meant, roughly, that any enumeration of its members was bound to be endless.

Were there any such sets though?

One intuition, aired in the Introduction, in the discussion of the paradoxes of the one and the many, was that there could not be. But Bolzano argued that there were. The set of points in space, the set of points in time, and the set of natural numbers were all perfectly good examples. He also had an argument that the set of truths was infinite, which can be recast as follows.

Bolzano's argument: One truth is the proposition that Plato was Greek. Call this p_1. But then there is another truth p_2, namely the propostion that p_1 is true. And so on *ad infinitum*. Thus the set of truths is infinite.

But could such sets be treated as finished and determinate wholes? That is, could they be treated in such a way as to license the actual mathematical infinite?

Bolzano argued that they could. For, *pace* Aristotle, their infinitude, so far from having to be given over time, had nothing to do with time at all. Endlessness here was not to be construed temporally. Or alternatively, insofar as endlessness *was* a temporal notion, infinitude was not to be understood in terms of it. (Friendliness towards the actual mathematical infinite was bound to call certain traditional views of mathematical infinitude into question.) Bolzano knew that this raised paradoxes. Indeed he believed that most mathematical paradoxes concerned the infinite. But he argued that they could be resolved. First, he said, we had to shrug off the idea that a set existed only if we could think about collecting its members together. There was no reason to suppose that what existed was at all constrained by what we could think about. Secondly, concerning the paradoxes of the infinitely big, he said that these could be dispelled in the (Gregorian) way outlined above. They seemed threatening only when we illegitimately thought about them in finite terms.

Dedekind, whose work on continuity was aired in Chapter 4, §2, also took this line (somewhat later). In his case, far from being fazed by the paradoxes of the infinitely big, he welcomed them as providing a valuable insight into the nature of the infinite. For he thought that we could capitalize on the fact that the members of an infinite set could always be paired off with the members of one of its proper subsets,[5] whereas this could never happen in the case of a finite set, by *defining* an infinite set as a set of which this was true. He then had a Bolzano-like proof that at least one infinite set exists, in his case the set of thoughts.

> *Dedekind's argument*: Given some arbitrary thought s_1, there is a separate thought s_2, namely that s_1 can be an object of thought. And so on *ad infinitum*. Thus the set of thoughts is infinite.[6]

Bolzano's stance, however, was even more radical, because he argued that it was possible to calculate with infinite quantities. He made some rather fumbling attempts to show how, later to be much refined and improved upon by Cantor. One interesting feature of his discussion, which relates back to what was said above, was that he seemed to take it for granted that there were 'true' criteria for comparing sets in size, and that these were the 'subset' criteria. On Bolzano's view there just *were* fewer even natural numbers than natural numbers altogether, irrespective of the fact that they could be paired off.[7]

2 Turn-of-the-century work on the foundations of mathematics[8]

Around the turn of the century mathematicians were becoming increasingly self-conscious about their own discipline and about the need for

clarity and rigour in its foundations. This was brought on largely by what
had been going on in the calculus (see above, Chapter 4, §§2 and 3). There
was a growing feeling that mathematical self-respect would not be satisfied
until a secure grounding of this kind had been provided right across the
board. It seemed necessary to lay bare the basic principles on which
various branches of mathematics rested, and then to firm those principles
by setting them in a broader and more secure mathematical context.
There were two great achievements in the history of mathematics (over
and above those of peculiar reference to the calculus) which inevitably
acted as paradigms here: Euclid's axiomatization of geometry; and Descartes'
reduction of that geometry to analysis. Both paradigms exerted considerable influence. Dedekind effected a kind of Cartesian reduction of analysis
to arithmetic. He also set down some basic principles of arithmetic that
were later taken up and sharpened by Peano, who presented them as an
axiomatic basis for arithmetic in Euclidean style.[9]

The most important figure in this connection, however, was Gottlob
Frege (1848–1925), a German mathematician and, by common consent,
the greatest logician of all time. Frege was one of the first mathematicians
seriously to express the hope that mathematics could be made secure in this
way, and at the same time to make a significant contribution towards
fulfilling the hope. His declared ambition – and it was a grand enough
ambition – was to follow out a programme of the kind sketched above for
virtually all of extant mathematics, by reducing it to certain fundamental
and self-evident principles of logic. (I say 'virtually all' because he did not
in fact believe that this could be done for geometry. He saw Descartes'
achievement in a somewhat different light.) Having isolated some suitable
logical principles, he wanted to couch them in a sufficiently precise
language (something like Leibniz' '*Characteristica Universalis*' (see above,
Chapter 4, §2)) to enable all the relevant mathematical theorems to be
derived from them by rigorous, mechanical, step-by-step procedures,
without appeal to intuition.[10]

In the course of following out the programme, Frege combined some of
his own insights with insights that he shared with the likes of Bolzano,
Dedekind, and Cantor to help to vindicate the new heterodoxy that
mathematical study of the infinite was as capable as any other branch of
mathematics of being put on a firm footing. Indeed he, like Cantor, talked
in terms of infinite number. We shall see later how such talk was justified.
But one interesting feature of it was that it involved both Frege and Cantor
in exactly the opposite intuition to Bolzano's about the 'true' criteria for
comparing sets in size (though neither Frege nor Cantor put it quite like
that). They both adopted the 'correlation' criteria, whereby there were as
many even natural numbers as natural numbers altogether, irrespective of
the fact that the natural numbers included odd numbers besides.[11] In their
defence, and against Bolzano, it is certainly their approach that has

proved the more elegant, the more fruitful, and the more serviceable. It also captures the rough intuition that a set's size is – to put it very metaphorically – a matter of what you can still see when you back off from the set to such an extent that you can still see its members but can no longer make out what they are. Cantor gave voice to this intuition when he talked in terms of arriving at a set's size by 'abstracting' from its members.[12] (It is also interesting to note that Hume had much earlier endorsed a 'correlation' criterion, though not with its application to infinite sets in mind.[13])

In order to deal not only with infinite numbers but with the natural numbers too, Frege needed some guarantee that infinitely many things existed. For it was part of his view that for any natural number to exist there had to be that many things to be counted. Furthermore, the guarantee had to be purely logical, so as to accord with his proposed reduction of arithmetic to logic. He could not appeal to propositions or thoughts as Bolzano and Dedekind had done. So what could he do?

He made pivotal use of the concept of a set (bearing out what I said in the last section about the increasing importance of this concept). Indeed how far his project could be viewed as a success was going to depend largely on how far this concept could be regarded as a purely logical one. But if it could, Frege seemed to have the guarantee he required – witness the following argument. (This was not in fact Frege's own argument, but it is very similar to his.)[14]

The Fregean argument: Whatever else exists, there must at least be the empty set, the set with no members.[15] Then there is the set whose only member is the empty set. Then there is the set whose only members are these two. And so on *ad infinitum*. Thus there are infinitely many things.

However, a devastating problem lay in store for Frege. It turned out that the idea of a set harboured a deep paradox. This paradox was discovered and communicated to him by the great English philosopher and mathematician Bertrand Russell (1872–1970).[16] It is the first of the technical paradoxes of the one and the many that we shall encounter, and it runs as follows.

Russell's paradox: A set does not typically belong to itself. For example, the set of planets is not itself a planet, so it does not belong to itself; and the set of people is not itself a person, so it too does not belong to itself. On the other hand, it seems that some sets do belong to themselves: the set of sets and the set of abstract objects look like obvious examples. Consider now the set of all those sets that do *not* belong to themselves, which we can call R. Does R belong to itself? Well, any set X belongs to R if and only if X does not belong to X. So, in particular, R belongs to R if and only if R does not belong to R – a blatant contradiction.

Frege learned of this paradox after he had published the first fruits of his endeavours and just as he was in the throes of publishing more. It seemed to show that the concept of a set, far from being purely logical, was radically incoherent, at least as he had been understanding it. But the fact that this undermined the argument that infinitely many things exist was not the worst of it. The concept of a set had been a main pivot of his project. The paradox revealed a fundamental contradiction at the very heart of his system. It dealt a death-blow to his entire project. Frege died embittered, convinced that his life's work had been for the most part a failure.[17]

The problem was not only his, though. The fundamental role that the concept of a set had begun to play meant that there was now a crisis at the very heart of mathematics, just when mathematicians had been doing something to put it on a firmer footing. Moreover, the crisis seemed to have something to do with the infinite. (The set R in Russell's paradox would have to be infinite.) There had already been intuitive resistance to the idea that there could be infinite sets. Perhaps that resistance had now been vindicated. Certainly many of those who were still prepared to accept infinite sets, including, as we shall see, Cantor, nevertheless reacted to this paradox, and to another that we shall be encountering shortly, by disavowing sets that were not in some sense manageably small; although a set could be infinite, its infinitude had to be somehow mitigated (see below, §5).

Meanwhile Russell, who had independently conceived a programme very like Frege's and arrived at very similar results, valiantly carried on with it. In collaboration with his former tutor Whitehead, he published a monumental three-volume work *Principia Mathematica*[18] in which the aim was to show, just as Frege had aimed to show, that mathematics could be reduced to logic and given a secure axiomatic grounding. But of course, the concept of a set now had to be handled with greater care. Russell argued that a set was a different *type* of thing from its members (so there was no sense even in asking whether a given set belonged to itself); and that his paradox did not arise if a careful check was kept on this. I shall not dwell on the details of this view. But when he spelt it out, it had a particularly interesting consequence for the overall programme. He now not only required the existence of infinitely many things, just as Frege had done, he required the existence of infinitely many things all of the same type. And the details of how this came about meant that he was deprived of a Frege-like proof that the requirement was met. (It was not just that Frege's own proof had to be rejected. Russell had no satisfactory way of adapting it either. He also, incidentally, took issue with the kind of proof that Bolzano and Dedekind had invoked.[19]) The existence of infinitely many things therefore had to be taken for granted as one of the basic axioms, the so-called axiom of infinity.[20] And now it seemed impossible to justify the claim that all the axioms were self-evident principles of logic, as originally intended.[21]

Russell himself acknowledged this breakdown in the programme. But however that may be, he was happy to place himself firmly in the new tradition according to which the mathematically infinite was a perfectly respectable object of study. Impressed by recent technical innovations, above all Cantor's, he pooh-poohed Bergson's suggestion that continuity was not susceptible of discursive analysis (see above, Chapter 7, §3), and petulantly dismissed the idea that the mathematically infinite was a 'false' infinite, a vain attempt to reach 'true' infinity. He did not mention Hegel and his followers by name, but it was clear whom he had in mind.[22]

The lesson so far is curiously ambivalent. Just as the paradoxes of the infinitely big were being overriden and friendliness towards the actual mathematical infinite was looking more and more creditable, it seemed that the infinite was intimately associated with a new family of paradoxes, those of the one and the many; and yet such advances were being made that the friendship still seemed worth consolidating. These advances were being made above all by Cantor, who founded what is now recognized as transfinite mathematics. It was not that he was unaware of the new paradoxes. But, showing tremendous fortitude and foresight, he took them in his stride, and refused to let them undermine his work. We shall see shortly how.

What, from a mathematical point of view, was motivating Cantor? He was certainly suspicious of the Aristotelian orthodoxy that the mathematically infinite had to be potential. After all, if the existence of a never-ending process could be acknowledged at all, it could be acknowledged *now* (however literally or metaphorically these temporal expressions were to be taken). The domain of things involved in the process could be given in advance, as an actually infinite totality. 'Each potential infinite,' he wrote, '... presupposes an actual infinite.'[23] (This argument had already been mooted by Pascal – though in his case to support belief in an actual infinite that was beyond our comprehension, something that demanded faith rather than mathematical investigation.[24]) There was, however, more to it than that. Cantor was also responding to the pressures of his own early work, which had been concerned with real numbers and continuity, much like Dedekind's work in the same area that we looked at in Chapter 4. He had no truck with infinitesimals. He was adamant throughout his life that the whole idea of an infinitesimal was demonstrably inconsistent.[25] But he did become convinced that mathematicians needed to deal with infinite sets, such as the set of natural numbers. And, like Bolzano and Dedekind, he saw nothing to cause alarm in the traditional stumbling-block to this – the paradoxes of the infinitely big. He proceeded to spell out, with meticulous care, how infinite sets were to be treated. And thus he built up his most effective and powerful riposte to the weight of tradition and the thing that convinced him that he was in the right, namely a coherent, systematic, and precise theory of the infinite that he could just lay before any sceptical gaze. It is to this theory that we now turn.

3 The main elements of Cantor's theory. Its early reception[26]

One of the cruxes of the theory was that it rested on the 'correlation' criteria for comparisons of size. We have seen enough by now to take any paradoxical sting out of these. Indeed it may turn out, for all we know so far, that we can use such criteria to show that all infinite sets are the same size; and that is actually in line with one of our inchoate intuitions concerning the infinite. From now on we shall follow Cantor and use such phrases as 'same size' and 'bigger than' in their 'correlation' senses. Thus we shall say that two sets are the same size if their members can be paired off. And we shall say that one set A is bigger than another set B if the members of B can be paired off with some of the members of A but not with all of them. What the paradoxes of the infinitely big show, in these terms, is that all sorts of infinite sets *are* the same size. For example, the set of natural numbers, the set of even natural numbers, and the set of rational numbers are all the same size. There are many more results of this kind, as Cantor helped to establish.

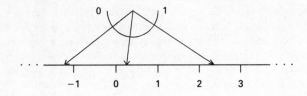

Figure 8.1

Consider the real numbers between 0 and 1 (exclusive). We can think of these as the points on a line-segment. By, so to speak, bending these into a semi-circle, we can establish that there are as many of them as there are real numbers altogether, as Figure 8.1 testifies. Cantor proved, in addition, that there are as many points on a line-segment, however short, as in a square, however big, or indeed as in the whole of space (where such points are all similarly construed in terms of real numbers). This was something that he personally found very unsettling, having spent three years or so trying to prove the exact opposite. He was famously prompted to write, in a letter to Dedekind: 'I see it, but I do not believe it'.[27]

So, *are* all infinite sets the same size? (Gregory for one thought so (see above, Chapter 3, §3).) Cantor was able to prove that they are not. All

```
0 — 0 · 3 3 3 3 · · ·
1 — 0 · 1 4 1 5 · · ·
2 — 0 · 4 1 4 2 · · ·
3 — 0 · 5 0 0 0 · · ·
    ⋮           ⋮
```

Figure 8.2

infinite sets are big, but some, it transpires, are bigger than others. It is this which gives the formal treatment of the infinite its bite. Were it not for the existence of different infinite sizes, transfinite mathematics would be the merest shadow of the subject it in fact is.

Again Cantor's proof ran directly counter to what he had previously spent some time trying to prove. He showed that there were more real numbers between 0 and 1 than there were natural numbers. He did this by arguing that any attempt to pair them off with the natural numbers was bound to result in failure; at least one such real was bound to be left over, not paired off with any natural number. His argument ran as follows.

Cantor's diagonal argument: Each real number between 0 and 1 can be expressed by means of an infinite decimal expansion, beginning with a 0 and a decimal point; this allows for the case of an expansion that terminates with a never-ending sequence of 0s. Thus

$$\tfrac{1}{3} = 0.3333\ldots;$$

$$\pi - 3 \text{ (that is, the decimal part of } \pi) = 0.1415\ldots;$$

$$\sqrt{2} - 1 \text{ (that is, the decimal part of } \sqrt{2}) = 0.4142\ldots;$$

and

$$\tfrac{1}{2} = 0.5000\ldots.$$

Now consider *any* pairing off of the natural numbers with a selection of these reals. For the sake of argument let us suppose that 0, 1, 2, and 3 have been paired off with the four reals just mentioned, in that order. Such a pairing off can be represented as an 'infinite square', as shown in Figure 8.2. Suppose next that we start with the first digit in the decimal expansion of the first real in this 'square', then move to the second digit

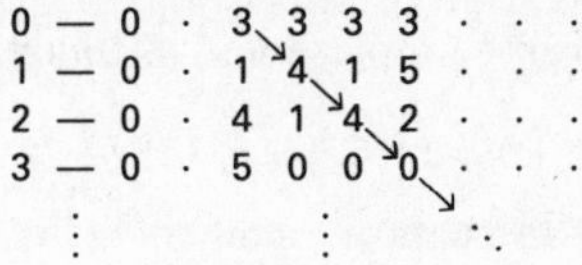

```
0 — 0 · 3 3 3 3 · · ·
1 — 0 · 1 4 1 5 · · ·
2 — 0 · 4 1 4 2 · · ·
3 — 0 · 5 0 0 0 · · ·
    ⋮           ⋮
```

Figure 8.3

119

in the decimal expansion of the second real, and so on indefinitely down the 'square's diagonal', as shown in Figure 8.3. If, each time we arrive at a new digit, we write down a 3 if that digit is a 4, and a 4 if that digit is anything other than a 4, then we shall find ourselves writing down what may itself be regarded as the decimal expansion of a real number between 0 and 1. In this particular example the expansion will begin with a 4, two 3s, and a 4. That is, the resultant real will be 0.4334 .... Now, has *this* real been paired off with any natural number? That is, is it itself one of the reals listed in the 'square'? No. It has been so defined that it differs from the first of the reals in its first decimal place, from the second in its second, from the third in its third, and so on *ad infinitum*. It cannot itself be any one of them. What this shows is that, *regardless* of what pairing off we start out with, at least one real must inevitably be passed over. It will always be possible to define such a real by means of this kind of 'diagonalization'. The natural numbers cannot after all be paired off with the real numbers between 0 and 1. There are more of the latter. The set of real numbers between 0 and 1, and *a fortiori* the set of all real numbers, is bigger than the set of natural numbers.

Nomenclature: There will be many future references to Cantor's diagonal argument, and to the sets involved. It is therefore worth giving those sets names. Let us refer to the set of natural numbers as ℕ and to the set of real numbers as ℝ. Then what the diagonal argument shows is that ℝ is bigger than ℕ.

Comment: There is a pleasant historical quirk here. Just as studying a diagonal had led the Pythagoreans to realize that real numbers enjoyed a kind of abundance beyond the grasp of the natural numbers (see above, Chapter 1, §2), so too, in a different way, with Cantor.

Cantor's diagonalization is not, however, confined in its application to real numbers. The very same technique can be used to show that there are more sets of natural numbers than natural numbers themselves. For any such set can be represented by means of an infinite sequence of *yeses* and *noes*, registering whether successive natural numbers do or do not belong to it. The set of even natural numbers, for example, can be represented as follows:

⟨*yes, no, yes, no, yes, no,* ...⟩.

And the set of primes can be represented as follows:

⟨*no, no, yes, yes, no, yes,* ...⟩.

Thus any pairing off of the natural numbers with a selection of such sets can be represented as an 'infinite square' of *yeses* and *noes*. An arbitrary example is given in Figure 8.4. Moving down the 'diagonal', we can write down the sequence of a set that must inevitably fail to be listed in this 'square'. For we can write down a *yes* each time we arrive at a *no*, and a *no* each time we arrive at a *yes*. In this case we write down

<no, yes, yes, ...>.

This shows that N is smaller than its own power-set. (The power-set of a set *A* is the set of sets of members of *A*.) Moreover the argument can be generalized to show that any set is smaller than its own power-set. Thus there is no limit to how big an infinite set may be. N is infinite. Its power-set is bigger. *Its* power-set is bigger still. And so on *ad infinitum*.

		0	1	2	.	.	.
0	—	yes	no	yes	.	.	.
1	—	no	no	yes	.	.	.
2	—	no	no	no	.	.	.
⋮		⋮					

Figure 8.4

Cantor proved most of these results in his late twenties and early thirties (though not – then – by using the technique of diagonalization; that was something he devised later). The general tenor of the period, with its new-found indulgence towards the actual mathematical infinite, was reflected in the fact that many received his work with enthusiasm. Frege was broadly sympathetic, Dedekind and Russell very much so. And the French philosopher Louis Couturat (1868–1914) argued that Cantor's work was indispensable to the study of continuity, and would eventually be regarded with the same mathematical equanimity with which the study of negative numbers, irrational numbers, and the like was now regarded.[28]

But Cantor's work was by no means universally acclaimed. Two and a half millennia of mistrust and suspicion are not overcome as easily as that. For example, his teacher, the German mathematician Leopold Kronecker (1823–1891), displayed life-long hostility to his work and acrimoniously opposed him in a variety of ways, for example by trying to block his publications.[29] Kronecker had a kind of Pythagorean belief that the natural numbers were the only 'real' mathematical objects, and any mathematics that was not ultimately concerned with them was, like Cantor's work, so much 'mathematical nonsense'. 'God made the integers;' he famously said, 'all the rest is the work of man.'[30]

Again, the French mathematician Henri Poincaré (1854–1912) described Cantor's work as 'a perverse pathological illness that would one day be cured.'[31] He challenged Cantor's claim to have proved that ℝ was bigger than N. Cantor's proof could just as well be taken to establish merely that we could not *devise* a way of pairing off the the natural numbers with the real numbers, or indeed that ℝ was not a genuine set at all – presumably because the real numbers were somehow too unwieldy to be grouped together into one determinate totality.[32]

This, incidentally, was something urged by the American philosopher and mathematician C.S. Peirce (1839–1914). He had independently discovered that there was no way of pairing off the natural numbers with the real numbers, but he concluded that ℝ did not exist as a completed whole. At most it existed as something potentially infinite. However many reals had been actualized, there were always more waiting to be. A continuum, he felt, was precisely *not* just a set of points. It was something absolute, consisting of unactualized possibilities, cemented together in a way that defied description but of which we were aware in experience. This ties in with various currents of thought noted in the last chapter, in connection with the 'metaphysically small'.[33]

Overall Cantor was very depressed by how his work was received. He never shrugged off the conviction that it had been unjustly ignored and misconstrued. But he persevered. The first of his two major publications, which appeared when he was almost forty, pulled together many of these results, introduced the very important idea of an ordinal, which we shall be looking at in the next section, and set down some of the basic principles of set theory.[34]

There was, however, one question in particular that he could not answer, and it caused him a great deal of anguish. The question was essentially this.

Cantor's unanswered question: Given that ℝ is bigger than ℕ, how much bigger? Somewhat more precisely: are there any sets that are intermediate in size between them, or is ℝ the 'next infinite size up' from ℕ?

This question will be sharpened and its significance spelt out in Part Two. Cantor agonized long and hard over it, alternately thinking that he had established one answer and then thinking that he had established the other. He never finally settled it, a fact which did much to compound his life-long depression and despair over his own work.

All the same, twelve and fourteen years later, in two parts, came his second major publication, in which these ideas were developed, precise methods were introduced for measuring how big infinite sets were, ways of calculating with these measures were established, and transfinite mathematics was at last set on a firm footing.[35] (We shall be looking at some of this material in Part Two. The diagonal argument was introduced a few years before this.) No longer, it seemed, could the mathematically infinite be thought of as that which is immeasurable or that which is greater than any assignable quantity.

But the matter was not so straightforward.

The Mathematics of the Infinite, and the Impact of Cantor

4 The theory of ordinals. The Burali-Forti paradox

It was about this time that the paradoxes of the one and the many were coming to light, in such a way as to suggest that two and a half millennia of suspicion about treating the mathematically infinite as a measurable whole had not been so misplaced after all. One of the purest of these was the Burali-Forti paradox. This was first published by the Italian mathematician Cesare Burali-Forti (1861–1931), in 1897, the year the second volume of Cantor's work came out.[36]

Before it can be presented, we need to look at Cantor's theory of ordinals, a further key element in his overall theory of the transfinite. Cantor was not only interested in how *big* infinite sets were. He was also interested in ways of imposing order on to them. A set's size has nothing to do with any particular ordering of its members. So this allowed for various new concepts to be introduced. The theory of ordinals was his attempt to systematize this part of his enquiry.

Before the idea of an ordinal can be defined, we need to look at one particular kind of imposition of order, namely a well-ordering.

Definition of a well-ordering: A *well-ordering* of a set X (finite or infinite) is an imposition of order on the members of X satisfying the following conditions: it singles out one of the members of X as the first, unless, of course, X has no members (that is, unless X is the empty set); it singles out another member of X as the second, unless X has only one member; it singles out another as the third, unless X has only two members; and quite generally, it singles out, for each member of X that has already been singled out, another as its immediate successor, unless there are none left; more generally still, it singles out, for each *set* of members of X that have already been singled out (finite or infinite), a first to succeed them all, again unless there are none left.

We can gain a feel for this definition by considering various impositions of order that are *not* well-orderings. For example, the standard imposition of order on the whole numbers

$$\langle \ldots, -2, -1, 0, 1, 2, \ldots \rangle$$

is not a well-ordering because it does not single out one of them as the first. Again, the standard imposition of order on the non-negative rationals

$$\langle 0, \ldots, \tfrac{1}{4}, \ldots, \tfrac{1}{2}, \ldots, 1, \ldots, 1\tfrac{1}{2}, \ldots, 2, \ldots \rangle$$

is not a well-ordering because, although it singles out one of them as the first, it does not single out one of them as the second; it does not single out any rational as the immediate successor of 0. Finally, the *non*-standard imposition of order on the whole numbers which singles out all the natural numbers, in their standard order, before all the negative whole numbers, in their standard order

123

⟨0, 1, 2, . . ., . . ., −3, −2, −1⟩

is not a well-ordering: it does not single out any whole number as the first to succeed all the natural numbers.

On the other hand the standard imposition of order on ℕ

⟨0, 1, 2, . . .⟩

is a well-ordering. So too the non-standard imposition of order on ℕ which singles out all the positive whole numbers, in their standard order, before 0

⟨1, 2, 3, . . ., 0⟩

is a well-ordering. And the non-standard imposition of order on ℕ which singles out all the even natural numbers, in their standard order, before all the odd natural numbers, in their standard order

⟨0, 2, 4, . . ., 1, 3, 5, . . .⟩

is a well-ordering.

Now a well-ordering of a set X and a well-ordering of a set Y may have the same shape. That is, the two well-orderings may look the same when we prescind from the things actually being ordered; they may look the same if we back off from them to such an extent that we can no longer make out what the things actually being ordered are. (This is reminiscent of Cantor's idea of taking an *un*ordered set and abstracting its size from it (see above, §2).) For example, consider once again the first non-standard well-ordering of ℕ mentioned above:

⟨1, 2, 3, . . ., 0⟩.

There is an analogous well-ordering of just the odd natural numbers that has the same shape:

⟨3, 5, 7, . . ., 1⟩.

This shape is that of an infinite progression followed by a single thing. (It is not only non-standard well-orderings that have this shape. Consider the following standard well-ordering of these rationals:

⟨½, ¾, ⅞, . . ., 1⟩.)

Intuitively, the shape of a well-ordering is a matter of how long it is. A slightly longer well-ordering, with a slightly different shape, would be like the first of these but with 3 (say) at the end:

⟨1, 2, 4, 5, . . ., 0, 3⟩.

Ordinals (or *ordinal numbers*) are used to measure length, or shape, in this intuitive sense. Given any possible shape that a well-ordering might have, there is an ordinal which acts as a measure of that shape. We can thus say what shape a well-ordering has – how long it is – by specifying the relevant ordinal.

How do the ordinals discharge this function? *Via* a well-ordering of themselves. There is an endless supply of ordinals (ensuring the existence of an ordinal to act as a measure of every possible length or shape), such that:

(i) one ordinal is the first;
(ii) for each ordinal, there is another ordinal which is its immediate successor;
(iii) for each set of ordinals (finite or infinite), there is another ordinal which is the first to succeed them all.

(Conditions (i) and (ii) are in fact just special cases of condition (iii): think about the empty set, and the set whose only member is a given ordinal, respectively. Thus only condition (iii) strictly needs to be cited.)

Each ordinal acts as a measure of the shape of the well-ordering of its predecessors. It, so to speak, 'looks down' on its predecessors and acts as a measure of how long they (collectively) are.

Note that, if the ordinals are to discharge their function, it is essential that condition (iii) be satisfied. For given any set of ordinals, there must be another that looks down on all of *them*, together with those that lie in between them, so that it can act as a measure of how long they (collectively) are.

But what *are* the ordinals?

The first ones are identical with the natural numbers. More specifically, the first is 0 (this measures how long a well-ordering of the empty set must be), the second is 1, the third 2, and so on. But condition (iii) above states that for each set of ordinals, there is another ordinal which is the first to succeed them all. So there must be an ordinal which is the first to succeed all the natural numbers. It is referred to as ω.[37] (We must beware of thinking of ω as an extremely big natural number – so big that it lies infinitely far along the progression of natural numbers. This does not make sense. ω is not a natural number at all, but something of an altogether different kind.) Clearly ω acts as a measure of the length of the standard well-ordering of $\mathbb{N}$, the set of its predecessors. Given condition (ii), it must have an immediate successor. This is referred to as $\omega + 1$. (Note, however, that this terminology is a kind of suggestive metaphor, at least for the time being. Until its application to other ordinals has been explicitly defined, '+' only makes literal sense when applied to the natural numbers themselves. Similar remarks will apply subsequently, for example to my use of '×', the symbol for multiplication.) Clearly $\omega + 1$ acts as a measure of the length of the first non-standard well-ordering of $\mathbb{N}$ considered above:

$$\langle 1, 2, 3, \ldots, 0 \rangle.$$

The immediate successor of $\omega + 1$ is $\omega + 2$. This acts as a measure of the length of the third of the non-standard well-orderings of $\mathbb{N}$ considered above:

$$\langle 1, 2, 4, 5, \ldots, 0, 3 \rangle.$$

The immediate successor of $\omega + 2$ is $\omega + 3$, and so on. The first ordinal to succeed all of these is $\omega + \omega$, or $\omega \times 2$. This acts as a measure of the length of the second of the non-standard well-orderings of N considered above:

$$\langle 0, 2, 4, \ldots, 1, 3, 5, \ldots \rangle,$$

whose shape is that of one infinite progression followed by another. There follow $(\omega \times 2) + 1$, acting as a measure of the length of

$$\langle 2, 4, 6, \ldots, 1, 3, 5, \ldots, 0 \rangle,$$

$(\omega \times 2) + 2$, acting as a measure of the length of

$$\langle 2, 4, 6, \ldots, 3, 5, 7, \ldots, 0, 1 \rangle,$$

and so on without end. 'Without end' is here understood in the relevant radical sense, as determined by condition (iii). Every set of ordinals is succeeded by another ordinal, and the process continues again. Figure 8.5, picturing some of the first ordinals in their standard order, helps to make this graphic. (Of course, since every set of ordinals is succeeded by another, there must be one which is the first to succeed all of *these* – as it were, the first ordinal not in the figure. It is referred to as ε_0.[38] Were we still in the business of stating paradoxes of the infinitely big, we could include the fact that the ordinals which *are* in the figure, despite their untold complexity, are no more numerous than the natural numbers, pictured right at the top. That is, there are just as many ordinals preceding ω as ε_0.)

We are now in a position to consider the Burali-Forti paradox.

The Burali-Forti paradox: The ordinals are well-defined mathematical objects, just as the natural numbers and real numbers are. They seem to constitute a perfectly determinate mathematical totality. Can we not therefore collect them together and consider the set of ordinals, say Ω[39] (just as we have been considering the set of natural numbers N, and the set of real numbers R)? Given that there *are* infinite sets, there can surely be no objection to this. But consider: if Ω does exist, then it is a set of ordinals like any other, and so, by condition (iii), there must be another ordinal which is the first to succeed all its members, thus contradicting its pretension to contain *all* the ordinals. It is as if the very endlessness of the ordinals precludes their being collected together into a set. No set can be 'big enough'. We can think of it like this. If Ω exists, then it has a well-ordering, so there must be an ordinal which acts as a measure of how long this well-ordering is. But obviously none of the ordinals *in* Ω is itself big enough to do this. So Ω cannot after all contain all the ordinals.

$0, 1, 2, \ldots,$

$\omega, \omega + 1, \omega + 2, \omega + 3, \ldots,$

$\omega \times 2, (\omega \times 2) + 1, (\omega \times 2) + 2, (\omega \times 2) + 3, \ldots,$

$\omega \times 3, (\omega \times 3) + 1, \ldots,$

$\omega \times 4, \ldots,$

$\omega \times 5, \ldots,$

.

$\omega^2, \omega^2 + 1, \omega^2 + 2, \ldots,$

$\omega^2 + \omega, \omega^2 + \omega + 1, \omega^2 + \omega + 2, \ldots,$

$\omega^2 + (\omega \times 2), \ldots,$

.

$\omega^2 \times 2,$

.

$\omega^2 \times 3, \ldots, \omega^2 \times 4, \ldots,$

$\omega^3, \ldots,$

$\omega^4, \ldots,$

$\omega^5, \ldots,$

.

$\omega^\omega,$

.

.

$\omega^{\omega^\omega}, \ldots, \omega^{\omega^{\omega^\omega}}, \ldots$

Figure 8.5

Burali-Forti himself drew from this paradox the conclusion that ordinals could not be the measures that Cantor had taken them to be; they could not themselves be well-ordered. But Cantor himself had already discovered the paradox two years earlier, and had viewed it in what would nowadays be regarded as a much more sophisticated light.

5 Cantor's attitude to the paradoxes

Cantor was untroubled by the Burali-Forti paradox. He took it to confirm something that he had already believed for some time anyway, namely that some totalities *were* immeasurably big, infinite to such a degree that they could not be assigned magnitudes in the way that, say, ℕ could. They were too big to be regarded as genuine sets at all. The totality of all ordinals was an example. There was no such set as Ω. And this was enough to dispel the paradox. Nor, by the same token, was there such a set as R, the set of sets not belonging to themselves; this was enough to dispel Russell's paradox, discovered a few years later. Cantor described these as 'inconsistent totalities'. He said they were 'many's too big to be regarded as one's'. He also described them as 'absolutely infinite'.[40]

The problem now was that it was hard to escape the feeling that, in some strange way, the whole issue had been brought back to square one. It was as if sets like ℕ, in being subjected to mathematical study by Cantor, had shown themselves to enjoy a kind of finitude, while the *truly* infinite had

127

continued to resist mathematical investigation. *It* still eluded our grasp, just as it always had. (There could not be any sets that were *genuinely* infinite.) It is important to realize that Cantor himself had spoken in these terms. He had used such phrases as 'the truly infinite' when talking about inconsistent totalities. In a comment that would not have been out of place in the last chapter, he said:

> The Absolute can only be acknowledged and admitted, never known, not even approximately.[41]

The problem can be made somewhat more graphic. Given that inconsistent totalities have been acknowledged, it is a very good question why Cantor's diagonal argument does not show just what Poincaré and Peirce suggested it showed, namely that ℝ is such a totality – rather than a set that is somehow bigger than ℕ. But if ℝ is such a totality, how can we be sure that ℕ is not one too? And if ℕ, why not all the other supposedly infinite sets? But if we go that far, then we really are back at square one (save, perhaps, that the paradoxes of the infinitely big have been replaced by the paradoxes of the one and the many).

Cantor did not want to return to such a sceptical view of the mathematically infinite of course; he felt he could acknowledge the existence of ℕ and ℝ as sets because there was no contradiction in doing so. But nor, by any means, did he want to dissociate himself entirely from earlier thinking about infinity. There is one exceedingly important fact about him that I have not yet mentioned. He was deeply religious. He believed he had a God-given gift to effect a mathematical study of the infinite and thereby, in a way, to vindicate certain cherished views about the divine against the charge of incoherence. Both when he was working on the infinite, and subsequently, when he had eventually been forced away from it by opposition and despair and was able to devote more time to religious studies and theology, he believed that he was in the hands of God. And this Absolute that had revealed itself in his own formal work, in a way that was so reminiscent of more traditional views of the infinite, was embraced by Cantor as a vital part of his conception of God. Despite the enormous revolutionary impact of his work, there was a deep sense in which, in the last analysis, it propagated a time-honoured tradition within the history of thought about infinity.[42]

6 Later development: axiomatization

Whatever else the paradoxes of the one and the many showed, it was clear that uncritical acceptance of the notion of a set was no longer acceptable – if indeed there was any such thing as *the* notion of a set. One possibility was that there was a family of related notions that had been illegitimately conflated. At any rate, our original intuitions needed to be straightened out. Russell, as we saw, had one way of doing this. But standard contemporary work in set theory presupposes a somewhat different straightening out, more powerful than Russell's and closer to what Cantor had in mind. The guiding idea is that the members of a set enjoy a kind of logical priority over the set itself. They exist 'first'. So no set can be a member of itself. (But *contra* Russell it is not technically *senseless* to say that any given set is a member of itself, just false.) It follows that R, in Russell's paradox, would be the set of all sets, if it existed. But it does not exist. Like Ω, it constitutes a Cantorian inconsistent totality, for ever beyond reach. This is how the paradoxes are avoided.[43]

This account has an intuitive appeal of its own. But it is very hazy. In any case, the original paradoxes must make us wary of trusting intuitive appeal. Many mathematicians therefore prefer not to think in these terms. They prefer to think about sets axiomatically instead, thus revealing once again the lure of the Euclidean paradigm. That is, they prefer to ratify their theorems by proving them from a given stock of fundamental principles serving as axioms. Though these principles are *in fact* intended to conform with the account sketched above, they can, for mathematical purposes, be treated as if they were supplied by an oracle. There are nine such principles which are nowadays used in this way.[44] Most of them were devised by the German mathematician Ernst Zermelo (1871–1953).[45] Other mathematicians later refined and supplemented his efforts, notably Frænkel, and the resultant system is usually referred to as Zermelo-Frænkel set theory, or just ZF. Some of the principles are very simple, for example that no two sets have exactly the same members (in other words, a set is determined by its members). Another is the principle that every set has a power-set. Another guarantees the existence of infinite sets.[46]

The twentieth century has, however, witnessed further crises in the foundations of mathematics, as a result of technical work which reveals inherent limitations in precisely this kind of axiomatic approach. Two logicians in particular, the Norwegian Thoralf Skolem (1887–1963) and the brilliant Austrian Kurt Gödel (1906–1978), are key figures here. For now I shall present only the barest outline of what they proved, since their work will be a major preoccupation of Part Two (see below, Chapters 11 and 12). What they proved, each in his different way, was that, insofar as what was being sought was a full and precise determination of just what sets were like, ZF was in certain crucial respects deficient; and not only that, any other possible axiomatic system of this kind would be deficient too. (So it was not just a question of tinkering with ZF to improve it.) What Skolem showed was that, if the axioms of a system such as ZF faithfully reflected *any* interpretation of the key words 'set', 'member', and the like, then they would faithfully reflect lots, some of them quite unlike what was intended.[47] What Gödel showed was that no such system would ever be strong enough to enable us to prove every truth about sets – unless it was inconsistent, in which case it would enable us to 'prove' anything whatsoever, true or false. We might have hoped that **ZF**, or something like it, would meet two *desiderata*: (i) to determine the whole truth about sets; and (ii) to determine nothing but the truth about sets. Gödel showed that this was impossible. Any axiomatic system that met one of these *desiderata* would do so at the price of not meeting the other. The Euclidean paradigm could not be extended to all branches of mathematics.[48]

Which *desideratum* does *ZF* fail to meet? Obviously we hope not (ii). We hope its theorems are at least all true, and its basic principles consistent; otherwise we have made no real progress from our original naïve view of sets. But how do we establish that **ZF** meets (ii), without referring back to the hazy intuitive account of sets sketched above, which may, for all we know, be just as awry as what it replaced? The sting in the Gödelian tail was that he developed his proof to show that no precise mathematical methods could be used to help here unless they involved a kind of question-begging. Resort to intuition seemed indispensable after all.[49]

CHAPTER 9

Reactions

No one shall be able to drive us from the paradise that Cantor has created for us. (David Hilbert)

I would say, 'I wouldn't dream of trying to drive anyone from this paradise.' I would do something quite different: I would try to show you that it is not a paradise – so that you'll leave of your own accord. I would say, 'You're welcome to this; just look about you.' ...

   (For if one person can see it as a paradise ..., why should not another see it as a joke?) (Ludwig Wittgenstein)

At one level, nobody could fail to be impressed by Cantor's work. It showed mathematical craftsmanship of the very highest calibre. But there was great room for debate about its significance. The infinite seemed at last to have been subjected to precise mathematical scrutiny. But perhaps Cantor had been indulging in technical flights of fancy. Perhaps, as I suggested in the last chapter, he had actually confirmed some of the most deeply entrenched prejudices about the infinite with his talk of inconsistent totalities.

   Certainly it was too much to expect that the infinite would now uncritically be taken on board as an object of mathematical enquiry in just the way that Cantor had represented it. There was some immediate opposition and hostility to his work, as we saw in the last chapter. Later thinkers reacted against it at a somewhat deeper level, recognizing the importance and merit of what they were reacting against but using this to think through some of the most fundamental questions about the nature of the infinite (and of mathematics). It is to their work, and related work, that we turn in this chapter.

1 Intuitionism[1]

The Dutch mathematician L. E. J. Brouwer (1881–1966) founded what was to become one of the most influential schools in the philosophy of mathematics, intuitionism. One of his main contentions was that mathematical statements had content: they were *about* something. Mathematics was not just a game in which meaningless symbols were manipulated in various ways. It had to confront mathematical experience, and answer to it. Picking up on an idea that had been present in both Aristotle and Kant, Brouwer argued that the experience to which all of mathematics answered (including geometry) was the experience which we each enjoyed of the pure structure of time. Starting with this, each of us was able to 'construct' the subject matter of mathematics. How so?

Time was a continuous (seamless) whole, but we could separate it into parts – past and future, say – in our thought. By doing this we arrived at what Brouwer called 'the basal intuition of mathematics', namely 'the intuition of the bare two-oneness'. (He also referred to this as 'the falling apart of moments of life into qualititively different parts, to be reunited only while remaining separated by time.'[2]) We could then indefinitely repeat this process, over time. We thus arrived at the fundamental idea of a progression. This in turn gave rise to two of the structural concepts integral to mathematics, infinitude by addition and infinitude by division.

This led Brouwer to re-embrace an Aristotelian conception of the infinite. The infinite was something that had to be given, or (better) constructed, over time. Its existence was potential, never actual.

Cantor's work had run directly counter to this. It had therefore, in Brouwer's view, lost touch with mathematical experience – the experience of the structure of time. Any interest it had was of a purely formal kind. It showed some of the technical tricks that could be played with (finite) mathematical symbols, but it did not truly relate to the infinite. For example, Cantor's principal defence for treating $\mathbb{R}$ as a completed whole was that it was consistent to do so. But for Brouwer, mere consistency did not make pronouncements true, or even meaningful. To have meaning, they needed to answer to mathematical experience. Brouwer could not accept that there was any such thing (completed whole) as $\mathbb{R}$; no such totality could be constructed from mathematical experience. What could be constructed were individual real numbers. This was done through the specification of *laws* determining their decimal expansions (say). But there was no way of constructing the real numbers *in toto*. (With $\mathbb{N}$ it was different. There was a single principle whereby successive natural numbers were constructed, that of adding 1. Talk of $\mathbb{N}$ as a whole could be legitimated by our grasp of this principle.) Brouwer's attitude to the diagonal argument was somewhat like Poincaré's (see above, Chapter 8, §3). It did not show that one thing was bigger than another. What it showed was that, given any progression of real numbers, there was a general recipe available for constructing a real number not in the progression.

If this critique is right, as many intuitionists still believe, then much of Cantor's edifice crumbles. But so too, significantly, does much of so-called classical mathematics, mathematics of the kind that has been accepted ever since antiquity. Why is this?

132

Consider the natural numbers. It is impossible for them all to be constructed within time. But we are inescapably immersed in time; and, for intuitionists, mathematical statements have to derive their meaning from what we, so immersed, are capable of constructing. How then is arithmetic possible? There is no problem with statements about particular natural numbers, for example that $7 + 5 = 12$. We can always construct a large enough finite selection of natural numbers to effect the relevant calculations. But there *is* a problem with (true) generalizations about the natural numbers, for example that each one is the sum of four squares (in the way that, say, $21 = 16 + 4 + 1 + 0$). The intuitionist solution to this problem is as follows. They argue that, in the case of a generalization of this kind, what we can construct is a proof, which trades on the principle whereby successive natural numbers are constructed (the principle of adding 1), and which enables us to see, in advance, why every natural number must have the property in question. For example, we might be able to see that 0 has it; and we might be able to see that, if some arbitrary natural number n has it, then $n + 1$ must have it too; we can then conclude, by the principle known as mathematical induction, that every natural number must have it. What is unintelligible on this view is the idea of an 'infinite co-incidence', whereby it so happens that every natural number has some property but there is no proof of this kind to show why. For this to be intelligible it would have to be possible to construct all of the natural numbers, in a finite time, and then to determine, by brute force, that they were alike in the relevant respect. But this is precisely what is not possible.

But why does this cast doubt on classical mathematics? This is best answered in terms of an example. Consider Goldbach's conjecture, that every even number greater than 2 is the sum of two primes (as, say, $8 = 5 + 3$). At the time of my writing this, no counterexample to the conjecture has been discovered: but neither has the conjecture been proved. Still, it is part and parcel of classical mathematics to assume that either there *is* a counterexample or the conjecture is true. In recoiling from the possibility of an infinite co-incidence, intuitionists cannot share this assumption. It is not that they have in mind a third alternative. Their position needs to be subtler than that. It is rather that they are not prepared to invoke the assumption, or others like it, in the course of their mathematical reasoning. Their Aristotelian conception of the infinite thus poses a challenge to some of the most elemental presuppositions of classical mathematics.[3]

2 Finitism[4]

If Brouwer had thrown down the gauntlet, then the great German mathematician David Hilbert (1862–1943) was prepared to pick it up. He famously described Cantor's creation as a 'paradise'.[5] Though well aware that it was not without its problems, he felt that it was worth every effort to defend, along with the framework of classical mathematics within which it had been presented. Given all its critics, and given also the newly discovered paradoxes, this was a daunting challenge. What was needed, Hilbert thought, was a careful critique of the infinite, something that would serve to explain and to justify just what Cantor had been up to when he represented the infinite as an object of mathematical study. A similar critique of infinitesimals had already shown how talk of the infinite was sometimes just a harmless *façon de parler*, a fact that had been recognized before by Leibniz and indeed by the medievals. Perhaps the same might prove true with respect to Cantor's work.

The first thing to note, Hilbert said, was that, given recent scientific developments, there was less and less empirical backing for the concept of infinity. The empirical evidence suggested – and still does – that the physical world was only finitely big and only finitely divisible. (Kant had been wrong to suppose that no empirical evidence could ever bear on this question (see above, Chapter 6, §3).) Likewise, perhaps, with space and time themselves.[6] For example, the possibility that space was finitely big had been mooted by the German physicist Albert Einstein (1879–1955) in the context of his cele- brated theory of relativity. The idea was that space might be finite but un- bounded, in something like the way in which the space of two-dimensional beings on the surface of a sphere would be finite but unbounded; finite, because any sufficiently long journey 'in a straight line' away from a given point would bring them back to it.[7] (The possibility of finite but unbounded space undercuts Archytas' inference from the universe's having no edge to its being spatially infinite (see above, Chapter 1, §5).)

It seemed, then, that any defence of the infinite, as construed by Cantor, needed to be internal to mathematics. The way to proceed, Hilbert argued, was to start with those finitary parts of mathematics that were incontrover- tible; then, renouncing all intuitions concerning the infinite (for by now, these were all inevitably suspect or question-begging), to see how talk about the infinite might nevertheless fit in. What were incontrovertible were simple numerical equations such as '7 + 5 = 12' and the like. Unlike Brouwer, Hilbert did not think that the meaning of such equations was to be accounted for in terms of our experience of time. Rather, such equations reported the results of combining and manipulating sequences of perceptible signs in various ways. To explain this, he envisaged a crude system of numerals whereby each positive natural number was represented by the same number of strokes. When such equations were spelt out in this symbolism, they very nearly *were* their own meaning. For example, instead of '7 + 5 = 12', we could write

$$||||||| + ||||| = ||||||||||||,$$

and then see by inspection that it was true. Equations of this kind Hilbert re- ferred to as *finitary* propositions. But there were also various generalizations concerning the natural numbers, references to ℕ as an infinite whole, and suchlike. These he referred to as *ideal* propositions. It was clear, in Hilbert's mind, that ideal propositions played a very important role in mathematics. His task now was to account for this.

We have already seen the account that Brouwer gave. But Hilbert wanted something that would do less violence to classical mathematics and indeed to Cantor's work. He knew he could not proceed by explaining the meaning of ideal propositions in terms of how they hooked up with some (infinite) feature of mathematical reality. To do that would have been to disregard the very problems that had prompted his enquiry in the first place. Instead he argued that ideal propostions *had* no meaning. They did not correspond to anything in mathematical reality in the way that finitary propositions did. Rather they were supplementary devices of a purely formal kind that were brought in to facilitate proofs and to make for greater elegance and perspicuity. Thus instead of proving in laborious detail that every natural number less than a million was the sum of four squares, we could take a short-cut *via* the ideal propostion, '*Every* natural number is the sum of four squares.' The concept of the infinite was therefore something like a Kantian Idea, an *a priori* concept applied beyond the realm of experience, to guide and regulate mathematical practice. There was no such thing as the infinite, which is why Hilbert's position is often called finitism, but we could proceed *as if* it existed. Hilbert himself put it as follows:

> Nowhere is the infinite realized; it is neither present in nature nor admissible as a foundation in our rational thinking. ... The role that remains to the infinite is, rather, merely that of an Idea – if, in accordance with Kant's words, we understand by an Idea a concept of reason that transcends all experience and through which the concrete is completed so as to form a totality.[8]

What Hilbert needed to do now was to show that this vindicated Cantor's transfinite mathematics, with all its classical presuppositions. Of course, since there was no question of the ideal mathematics having to be true, it could be rigged to precisely this end. But there needed to be some guarantee that we could not then use it to prove things that we did not want to prove (say, that $7 + 5 = 13$); otherwise it would have lost its rationale. *It* did not have to be true. But what *issued* from it did. However, Hilbert was confident that suitable ratification could be provided. He argued that this would involve carrying out a two-part programme, which subsequently became known as Hilbert's programme.

Hilbert's programme:
(i) The new mathematics, including all its ideal elements, would have to be formalized, by being cast axiomatically.

(ii) Its consistency (with finitary mathematics) would have to be established, without – question-beggingly – presupposing any of its ideal methods of proof.

(ii) would have to involve investigating the purely formal properties of these ideal methods in a way that would be tantamount to doing straightforward finitary mathematics. Completing the programme would have the added bonus of putting paid to any worries we might still have about the possibility of further paradoxes lurking, of the kind that had already come to light.

Hilbert presented all of these ideas in a paper in which he also sketched how the programme might be completed.[9] Towards the end of the paper he 'played a last trump'. He gave an outline of how the theory he had presented, also in outline, could be used to settle Cantor's unanswered question about the size of ℝ (see above, Chapter 8, §3). He attempted to show that ℝ *was* the next infinite size up from ℕ.

However, he used assumptions that very few later mathematicians were prepared to accept. Not that that was by any means the greatest setback to his efforts. Much more serious was Gödel's work, mentioned at the end of the last chapter. This appeared only six years later, and it seemed to show conclusively that in fact neither part of his programme could be satisfactorily carried out. I shall discuss this in further detail in Part Two (see below, Chapter 12, §3).

Prior to Gödel's discoveries, however, there were many who embraced Hilbert's finitism. The German philosopher Hermann Weyl (1885–1955), having at one stage sided with Brouwer, later came round to the view that we did have access to infinite wholes – in the sense licensed by Hilbert, namely that we could use symbols as if they stood for them, in ideal propositions.[10]

The Austrian philosopher Felix Kaufmann (1895–1949) likewise accepted much of Hilbert's critique. He denied that there were strictly any infinite sets, but insisted that we could talk about ℕ as a convenient *façon de parler*, for this made it easier to discuss finite sets. (It was like talking about the average parent: there was no such person but it made it easier to discuss real parents.) However, unlike Hilbert, Kaufmann did not feel that the whole of Cantor's transfinite mathematics could be vindicated in this way. For it was *too* far removed from reality, that is the reality of the natural numbers as described in simple equations such as '7 + 5 = 12'. (This was reminiscent of Kronecker (see above, Chapter 8, §3).) For example, like Brouwer, Kaufmann denied the existence of ℝ and took the diagonal argument to show merely how, given a recipe for filling in an 'infinite square' of digits, we could construct a real number whose decimal expansion was not on any row of the 'square'.[11]

Although these reactions pre-dated Gödel's work, Hilbert's ideas have continued to be of great interest, and to exert considerable influence in the philosophy of mathematics. There are those who think many of them are still viable. They were also an important spur to Wittgenstein.

3 Wittgenstein

Ludwig Wittgenstein (1889–1951) was an Austrian who spent much of his life in England. For me he stands with Heidegger as one of the two giants of twentieth-century philosophy. It is customary to divide his work into two phases, and, despite profound continuities between the earlier phase and the later phase, this is entirely apt. (It is said that one of the things that initiated the later phase, by reawakening his interest in philosophy after a long period of philosophical inactivity, was attending a lecture by Brouwer on the foundations of mathematics.) I shall be making special use of his earlier work in Part Two (see below, Chapter 13). But the main focus of this section is his later work, which included much on the topic of infinity.

One of the ideas that dominated his later work was that the meaning of a word was a matter of how it was used. Words were like tools. To grasp a concept was to be clear about the use of the words and phrases governing it. As soon as words were wrenched from their proper use and mishandled, (needless) philosophical perplexity arose. For it became possible to frame all sorts of pseudo-questions which posed as philosophical problems but which, in the nature of the case, we did not have the wherewithal to answer. For example, we could imagine mishandling a phrase like 'the average parent'. While it makes good sense to say, 'The average parent has 2.4 children,' it does not make sense to say, 'The average parent is expecting another child.' A philosophical 'problem' might arise about what would happen to all ordinary parents if the average parent found herself (himself? itself? theirselves?) expecting another child. Once we returned to the proper use of the phrase, and thus to a correct grasp of the concept, this 'problem' would be dissolved. Wittgenstein saw something similar in traditional discussion of the infinite.

He believed that the correct use of terms such as 'infinity' was to characterize the form of finite things and, relatedly, to generalize about the endless possibilities that finite things afford. (We shall see more clearly in a little while what this amounts to.) It was incorrect to apply such terms directly to what we encounter in experience. And it was incorrect to use them to describe anything as being actually infinite. So, for example, we could say that there were infinitely many numbers. But this must mean that however many numbers we had counted we could always count more (and not, so to speak, because there *was* no last number, but because the phrase 'last number' made no sense.[12]) Again, we could say that space and time were infinite. But this must mean that it was part of the form of a spatio-temporal object to have various unlimited possibilities of movement:

however far such an object had travelled, there would be space and time enough for it to travel still further. There was no question here of an 'infinite reality'.[13] So too we could say that space was infinitely divisible. But this must mean:

> Space isn't made up of individual things (parts) ... [It] gives to reality an infinite opportunity for division.[14]

(Of course, there were empirical issues here that were not to be prejudged. But Wittgenstein was only talking about what made *sense*.)

To describe something as actually infinite, then, was not just a mistake. It was a mishandling of the language. It was like saying, 'The average parent is two months pregnant,' or, 'It is 5 o'clock on the sun.'[15] And it was Wittgenstein's belief that once this fact was properly cognized, then philosophical perplexity about the infinite would at last be dissipated. A good example was the perplexity surrounding the paradox of the divided stick. For Wittgenstein, there was an incoherence in the very setting up of this paradox; no genuine situation had been described. It made sense to say, 'This stick is infinitely divisible.' It did not make sense to say, 'This stick is (has been) infinitely divided.'[16] Moreover, if we did say, 'This stick is infinitely divisible,' it was important that we should be clear as to precisely *what* sense it made. As we saw in our discussion of Aristotle, such sentences are ambiguous. To construe it in the wrong way (as meaning that a situation could be brought about in which this stick was divided into infinitely many pieces) was to start mishandling the language again, reopening the possibility of ill-begotten philosophical conundrums. What it meant was that however much the stick had been divided it could always be divided more.

Relatedly, we had to resist the idea that infinity was something like a natural number, only much bigger (so that, for example, three was somehow closer to infinity than two). For proper uses of 'infinity' were very different from proper uses of 'two' or 'three'. Even if it made sense to say that a path was infinitely long, it made a very different kind of sense from saying that the path was three miles long. An infinitely long path was, as Aristotle would have said, a path that could never be traversed – that is, a path with no end, not a path with an end infinitely far away. Again, an infinite set was a completely different *kind* of thing from a set with three members. 'Set' was hardly even univocal in the two cases.[17]

Many of Wittgenstein's conclusions were reminiscent of those of his great predecessors. Aristotle too would have found the wrong construal of 'This stick is infinitely divisible' unintelligible. Kant would have licensed various claims about the inifinite whole, but then insisted that we be clear as to what sense they made: they were to be understood as injunctions (involving a regulative use of our Idea of the infinite whole) and not as ordinary assertions. But the way Wittgenstein arrived at his conclusions, that is *via* careful scrutiny of the use of words, was in fact as reminiscent of the medievals as of anyone. It called to mind the categorematic/syncategorematic distinction, an item of essentially grammatical categorization. Wittgenstein himself was for ever talking about the 'grammar' of words, when appealing to their proper use. It was almost as if he was saying, or was committed to saying, that the 'grammar' of 'infinity' meant that it could be used only syncategorematically.

But what about its use in mathematical contexts? What about 'Cantor's paradise'?

Wittgenstein felt very strongly that it was not his business, as a philosopher, to interfere with mathematical practice.[18] His task was carefully to *observe* mathematical practice, to gain a clear view of how mathematical expressions were used, and then, where appropriate, to try to combat their misuse.

Did he then have to regard Cantor's work as sacrosanct, or at least as immune to philosophical criticism?

Not exactly. For one thing, mathematics was not a completely isolated discipline. Mathematical treatment of the infinite was not, and could not be, independent of how it was treated in other contexts. In any case, as we saw when looking at the early history of the calculus, it is not impossible for mathematicians themselves to mishandle their own apparatus and to import conceptual confusion into their own discipline. This raised a problem of circularity, though. How was Wittgenstein to know what to observe in the first place? Being able to distinguish between cases of legitimate mathematical practice and cases of mathematicians themselves going astray seemed to require the very discernment that was supposed to be acquired by observation of legitimate mathematical practice. If this circularity was not to be vicious, Wittgenstein did after all need to approach mathematics with a degree of humility. This, I think, helps to explain an ambivalence in his attitude to Cantor's work. On the one hand he felt pressure not to challenge it in any way. That, for better or worse, was what mathematics was now like. On the other hand his own views concerning the infinite meant that he did not like the tenor of the work *at all*. Still, there was a perfectly reasonable way out for him. What he did was to let the work stand but to remonstrate strongly against certain *attitudes* towards it. He was struck by Hilbert's view, shared by many, that Cantor had created a mathematical paradise. For Wittgenstein it could just as well be seen in a quite different light. 'Imagine set theory's having been created by a satirist,' he wrote, 'as a kind of parody on mathematics.'[19]

But it was not principally at Hilbert's attitude that Wittgenstein took umbrage. It was at an attitude quite foreign to Hilbert, though prevalent elsewhere, namely that transfinite mathematics served to describe a kind of super-physical landscape with all its bumps and nooks and crannies, populated by objects of various different sizes. This attitude was very much

Gödel's for example. Gödel did not express it exactly like that. But he did have a robust sense of mathematical reality. In a discussion of Cantor's unanswered question about the size of $\mathbb{R}$, the question that Hilbert took himself to have settled but only by a kind of *fiat*, Gödel said that we perceived mathematical objects in something like the way in which we perceived physical objects; certainly they were just as real, and Cantor's question was a genuine question about something quite independent of us, *not* a question to be settled by *fiat*.[20] This whole view of mathematics was an anathema to Wittgenstein.

Wittgenstein believed that when we scrutinized transfinite mathematics, what we saw was a variety of formal techniques, proof-procedures, and the like, but that, in a sense, was all there was to it. There was no 'landscape' being described. There were no 'objects' being perceived. And certain ways of couching the results were to be deplored insofar as they encouraged the idea that there were. 'The dangerous, deceptive thing about the idea: ['$\mathbb{R}$ is bigger than $\mathbb{N}$'] . . .,' he wrote, 'is that it makes the determination of a concept . . . look like a fact of nature.'[21] Similarly, he was highly suspicious of the kind of account of the relationship between real numbers and rational numbers that I gave above in Chapter 4, §2, whereby real numbers were seen as filling the 'gaps' between rationals.[22] By the same token, there was, for Wittgenstein, nothing mysterious or transcendent about transfinite mathematics, any more than there was about chess, or noughts and crosses. There was a kind of heady pleasure that we got from discovering that some infinite sets were bigger than others, like the pleasure of discovering that space was curved – what Wittgenstein would have called a 'schoolboy' pleasure.[23] This had to be resisted. There was nothing more to the diagonal proof than the technique actually set down on the page (that is, the technique for specifying a sequence of digits different from all those listed); and there was nothing more to the result than to the proof.

Uses of 'infinity' and related terms were quite *straightforward*. We had to learn to take them at face value. For instance, the three dots in

$$0, 1, 2, \ldots$$

were not an *abbreviation* for something too long to write down. They were themselves part of the mathematical symbolism with a perfectly precise, specifiable, unmysterious use.[24] The symbolism seemed puzzling and enigmatic only when we tried to look beyond it to what it was pointing to. It was not pointing to anything. *It* was the mathematical reality.

One very important consequence of these views was that Wittgenstein, like Brouwer, wanted to challenge the idea of an infinite co-incidence. Attention to the 'grammar' of generalizations about infinite totalities, based on close inspection of the relevant mathematical techniques and proof-procedures, revealed that the idea of an infinite co-incidence was unintelligible. What gave a true generalization about (say) ℕ its meaning was that we could recognize a proof of it. (We were seduced into thinking otherwise by the false picture of a determinate mathematical landscape out there, independent of what we could or could not prove about it.)[25]

It was here that Wittgenstein felt prepared to question standard mathematical practice – to see mathematicians as mishandling their own apparatus. For, like Brouwer, he believed that they made assumptions that they were not entitled to make once the idea of an infinite co-incidence had been rejected. They assumed, for example, that unless every natural number had a given mathematical property, there must be a counterexample.

As we can see, Wittgenstein's route to these intuitionist conclusions was quite different from Brouwer's. Something that helps to reinforce this point is the work of the English philosopher Michael Dummett (born 1925). Dummett has done as much as anybody to show how broadly Wittgensteinian considerations about language and meaning can sustain conclusions very like those arrived at by Brouwer, though in a way that runs directly counter to much of what Brouwer himself believed. This is an apt point at which to begin our discussion of current thought about the infinite.[26]

4 Current thought

The mathematical experience on which Brouwer laid so much emphasis was essentially private and incommunicable. Dummett, by contrast, has taken as his starting point the essential publicity and communicability of mathematical ideas, and has tied this in with the Wittgensteinian tenet that the meaning of a mathematical expression must ultimately be a matter of how it is used in mathematics, just as the power of a chess piece is a matter of how it is used in chess. Were the meaning not something open to public view in this way (for example, were it some range of occult images or indeed were it the kind of thing Brouwer took it to be), it would be impossible for anybody ever to have learned it or to demonstrate that they had done so; and this in turn would make communication impossible. (Brouwer thought that communication *was* impossible, in any ideal form.) It follows, in Dummett's view, that no mathematical sentence could be true in a way that essentially outstripped any capacity we had to recognize it or prove it to be such, for its meaning would not then be relevantly dependent on the kind of use to which we put it, or indeed could put it. So, as we saw with Wittgenstein, the idea of an infinite co-incidence comes under threat again. Although Dummett's argument is radically opposed to Brouwer's, he is taking something for granted that marks a vital point of contact between them: the fact that we are inescapably immersed in time. If we could somehow escape from time, and thereby survey a totality like ℕ in its entirety, the idea of an infinite co-incidence might not be so suspect. But we cannot. The upshot is that Dummett, like Brouwer, and like Wittgen-stein, has urged us to take a much more critical attitude to some of the most elemental presuppositions of classical mathematics.[27]

All three have placed emphasis on our capacities – what we are capable of constructing, what we are capable of recognizing as a proof, and the like. But consider: if our immersion in time is taken to place severe constraints on these, then why not similarly our physical limitations? After all, there is a perfectly good sense of 'could' in which none of us could construct, or survey, a finite segment of ℕ that was so big that it included more members than the number of atoms in the known universe, or more members than the number of milli-seconds that will have elapsed by the time the earth has been swallowed up by the sun. So cannot these arguments casting doubt on the idea of an infinite co-incidence be extended to cast analogous doubt on the idea of a truth concerning some sufficiently large natural number?

Wittgenstein may have thought that they could. He sometimes verged on a correspondingly extreme position.[28] Others have recently explored the position more or less sympathetically.[29] But it finds no place in either Brouwer or Dummett. Indeed Dummett has argued that it is incoherent.[30] The problem that Brouwer and Dummett both thereby face (of steering a middle course) has thus become a focus of much debate. Brouwer could insist that what matter are our capacities insofar as they relate to the pure structure of time, because it is that which ultimately furnishes mathematics with its content. Dummett, for his part, has only ever wanted to maintain an open-minded scepticism about classical mathematics. He may be at liberty to continue to do so. A different, but related, problem is this: if physical limitations are *not* relevant here, then can we not, after all, construct, or survey, the whole of ℕ in a finite time, by starting with 0, then dealing with 1 twice as quickly, then dealing with 2 twice as quickly as that, and so on *ad infinitum*? This question was raised by Russell, who famously declared that the impossibility of performing infinitely many tasks in a finite time was merely 'medical'.[31] I shall return to this issue in Part Two (see below, Chapter 14, §6).

One thing that it helps to show is that the same old puzzles and preoccupations are as relevant as they ever were to discussion of the infinite. A survey of the current literature reveals a continuing concern with all the perennials: the distinction between the actual infinite and the potential infinite; the relationship between the infinite and time; Zeno's paradoxes; the paradoxes of thought about the infinite; and so forth. But current debates have the advantage of being informed by recent empirical and mathematical discoveries – if it can be called an advantage: these discoveries have in many cases served only to exacerbate and to set in sharper relief some of the old problems, something that the new paradoxes of the one and the many illustrate only too well. (This is a point that I shall try to develop in Part Two.) There is perhaps also more emphasis now, in the wake of Wittgenstein, than there used to be on questions of meaning and linguistic understanding. But still the main concern is how to understand our own finitude and our relation to the infinite.

Two American writers in particular deserve mention for having produced fascinating work on the infinite that helps to bear this out: José Benardete (born 1928) and Rudy Rucker (born 1946). Both have offered enlightening surveys of the current state of the art.[32] Benardete, whose lively sense of the paradoxical nature of the infinite has graced our drama more than once – the paradox of the gods, for example, is due to him (see above, Introduction, §1) – has argued, as against Wittgenstein, that saying that there are infinitely many stars is much the same sort of thing as saying that there are a trillion. Rucker, whose discussion of the paradoxes of the one and the many is especially gripping, has followed a route from them, and from the paradoxes of thought about the infinite, to a kind of mysticism. Some of his conclusions are very close to those that I shall try to defend in Part Two.

A final twist comes in the work of another American, the philosopher and logician W. V. Quine (born 1908). He has argued that it is only because there are infinitely many things that we need to operate with the fundamental notion of a *thing* at all. For this notion is used principally in making generalizations, for example when we say that every*thing* is thus and so. But if there were only finitely many things, we could make such generalizations by spelling out, one by one, what each was like.[33] Of course, the 'could' here is interesting. In what sense *could* we do this? We are reminded of the way in which Brouwer and Dummett were prepared to prescind from our physical limitations where they were not prepared to prescind from our temporality. Likewise, it seems, Quine.

Our historical drama has now finished. Not one of its protagonists was prepared to accept the infinite unconditionally. There was always a *caveat* against some conceptual aberration with which it might be associated, conflated, or confused, whether this was the actual infinite (Aristotle), or the unconditioned physical whole (Kant), or the mathematical infinite (Hegel), or inconsistent totalities (Cantor), or the idea of a super-physical mathematical landscape (Wittgenstein). Each of them *might* simply have rejected the idea of the infinite itself as a conceptual aberration. But none of them did. There was a momentary insurrection among the British empiricists, but otherwise this never looked like a serious option. I submit that it still does not. How then are we to view the infinite?

Part Two

Infinity Assessed

The symphony is a musical epic. We might compare it to a journey leading through the boundless reaches of the external world, on and on, farther and farther. Variations also constitute a journey, but not through the external world. You recall Pascal's *pensée* about how man lives between the abyss of the infinitely large and the infinitely small. The journey of the variation form leads to that *second* infinity, the infinity of internal variety concealed in all things. . . .

Man knows he cannot embrace the universe with all its suns and stars. But he finds it unbearable to be condemned to lose the second infinity as well, the one so close, so nearly within reach. . . .

That the external infinity escapes us we accept with equanimity; the guilt over letting the second infinity escape us follows us to the grave. While pondering the infinity of the stars, we ignore the infinity of [the one we shall lose]. . . .

It is no wonder, then, that the variation form became the passion of the mature Beethoven, who . . . knew all too well that there is nothing more unbearable than losing a person we have loved – those [few] measures and the inner universe of their infinite possibilities.

(Milan Kundera)

CHAPTER 10

Transfinite Mathematics[1]

Thus quantum impels itself beyond itself; this other which it becomes is in the first place itself a quantum; but it is quantum as a limit which does not stay, but which impels itself beyond itself. The limit which again arises in this beyond is, therefore, one which simply sublates itself again and impels itself beyond to a further limit, *and so on to infinity*.

(G. W. F. Hegel)

It is often said that mathematics is the science of the infinite.[2] And yet, before the advent of Cantor's work at the end of the nineteenth century, few mathematicians even looked upon the infinite as a serious object of mathematical study. Many still do not. This situation is not as crazy as it sounds. Even when the infinite is not itself serving as an object of mathematical study, mathematicians can still be said to be exploring the infinite insofar as what they *are* studying presupposes an infinite framework. (This was a point that first arose when we were looking at early Greek mathematics (see above, Chapter 1, §5).) When the infinite does become an object of mathematical study, as in contemporary set theory (which is the modern development of Cantor's pioneering work on the infinite), it is as if mathematicians have chosen to step back and scrutinize the framework itself. If mathematics is the science of the infinite, then set theory is self-conscious mathematics.

My aim in the next three chapters is to look further into that self-conscious mathematics and to explore other technical work that bears directly on the infinite. This work will serve as a useful peg on which to hang a number of more general ideas. It will help to crystallize many of the puzzles and conundrums that beset any inquiry into the infinite. Later in Part Two the discussion will be extended to broader, non-mathematical issues.

1 The iterative conception of a set. The paradox of the Set of all Sets

In Chapter 8, §6, I sketched an intuitive picture of what sets are like, the picture which informs contemporary set theory. (The nine axioms of **ZF**

147

are supposed to be a faithful reflection of it.) It is also more or less the picture which Cantor himself had in mind. The conception of a set that underlies it is often referred to as the *iterative* conception.

In this section I wish to describe it in more detail. I shall do so in a way that is largely uncritical. It is not that I wish to ignore the various objections that we saw levelled against it in Chapters 8 and 9. The point is rather to present a platform for future discussion. Some of the objections will come to the fore again later.

It is most convenient, in fact, to confine attention to sets of a particular kind; for these can act as representatives of all the others. The sets in question are those that would still exist even if there were nothing *but* sets. (That is, they would still exist even if there were no people, planets, *et cetera*.) The empty set is an obvious example – as is the set whose sole member is the empty set. Such sets are often referred to as *pure* sets. I shall adopt the convention of using a capital 'S' and refer to them as Sets. Thus all Sets are sets, but not *vice versa* (the set of people is not a Set); and the members of a Set must themselves be Sets.

So, what Sets are there?

First, there is – to repeat – the empty set, which we can refer to as $\emptyset$.

Then there is the Set whose sole member is $\emptyset$, namely $\{\emptyset\}$. (Recall that the notation for the set whose members are x, y, z, . . . is '$\{x, y, z, . . .\}$'.[3])

Then there is the Set whose sole member is $\{\emptyset\}$, namely $\{\{\emptyset\}\}$, as well as the Set whose two members are each of the first two, namely $\{\emptyset, \{\emptyset\}\}$.

Sets are constructed stage by stage in this way. And they are constructed increasingly quickly.

At stage 1, there is only one Set, namely $\emptyset$.

By stage 2, there are two Sets.

By stage 3, there are four Sets.

In general, if the number of Sets constucted by stage n is k, then the number of Sets constructed by stage $n + 1$ is 2^k. This is because the Sets constructed by stage $n + 1$ are all the possible selections of Sets constructed by stage n. And for each of those k Sets there are just two possibilities: that it should be selected; or that it should not.

So by stage 4, the number of Sets is $2^4 = 16$.[4]

By stage 5, it is $2^{16} = 65,536$.

By stage 6, it is $2^{65,536}$ – a number that is so big it has almost twenty thousand digits. (To write it out in full would have required at least ten pages. By contrast the number of seconds which make up the five-thousand-million-year history of the earth has only eighteen digits.)

By stage 7, our minds begin to reel and crack under the strain. It is a curious fact that we cannot cope with a number this big whereas we can, apparently, cope with ω – the first ordinal to succeed all the natural numbers. However, this may be neither particularly significant nor, as one might also think, unduly suspicious. It may just mean that we need a careful account of how 'cope' is supposed to be understood here.

All the Sets that we have considered so far have been finite. But there exist infinite Sets too, for example the Set of all those Sets constructed by stages 1, 2, 3, . . . (These stages proceed *ad infinitum*.) And these infinite Sets themselves belong to further Sets. The process of Set construction never stops. In fact the stages by which Sets are constructed are endless in the same way that the ordinals are endless: given any set of them (finite or infinite), there is another that succeeds them all. Moreover, there is a first such. Their ordering (by precedence) is a well-ordering. Thus we can use the ordinals themselves to index the stages. After the finite stages 1, 2, 3, . . ., there is stage ω. Then there is stage ω + 1. And so on without end. The general principle is that for each ordinal α, there is a stage α.

Sets, we see, form a 'V'-shaped hierarchy. At its base there is Ø, and every other Set lies somewhere further up, constituting a collection of Sets taken from below it. The hierarchy has no top. Any attempt to close it off would abnegate the very idea of the endlessness of Set construction. They would 'burst through'. This ties in with a deep temporal metaphor that underlies the iterative conception (witness its name, and witness also the fact that it is so natural to talk in terms of 'Set construction'). According to this metaphor the members of a Set must exist *before* the Set itself, ready to be collected together, and the different stages by which Sets are constructed are different stages in time, so that a temporal axis can be thought of as running up the middle of the hierarchy: to say that Set construction is endless is to say that it is never (at any stage in time) complete. The infinitude of the Set hierarchy is thus potential, never actual. It is spread over endless time.[5] (This whole account smacks of intuitionism of course (see above, Chapter 9, §1); and indeed it has been argued that the hierarchy cannot properly be described except in intuitionistic terms.[6])

There is, according to this picture, no such thing as the Set of all Sets. No Set can contain Sets constructed at the same or later stages. In particular, of course, no Set can contain itself. Any Set must be merely one Set among others *within* the hierarchy, waiting to be followed in its construction by endlessly many others.

There is something very Kantian about this situation. Suppose we assume that the Set of all Sets (as it were, the unconditioned whole) *does* exist. Then we can construct the following antinomy.

(a) The Set of all Sets must belong to itself. For if it did not, it would not be the Set of *all* Sets; there would be at least one Set – itself – which it did not contain.

(b) The Set of all Sets must *not* belong to itself. For if it did, it would not be the *Set* of all Sets; a Set is something that cannot contain itself.

The (Kantian) solution to this antinomy is to drop the assumption. There is no such thing as the Set of all Sets. It neither belongs to itself nor fails to belong to itself. The concept of the Set of all Sets is at best an Idea of reason which can continue to have a legitimate regulative use – (perhaps we can use it to enjoin ourselves to carry on with our mathematical study of Sets) – but which does not pick out anything in mathematical reality.

This, however, is itself a paradox. It is another paradox of the one and the many, perhaps the purest of all. We must not forget that what we are engaged in now is supposed to be self-conscious mathematics. What then can prevent us from self-consciously reflecting on the framework within which the hierarchy itself lies? There have, after all, already been a number of references to the hierarchy, including the reference to it as being, according to the temporal metaphor, potentially infinite. This in itself seems to belie the idea that the metaphor can be taken at all literally. We have – *now* – mathematical access to the hierarchy as a whole. The Sets in it form a determinate totality about which we can make, and have been making, generalizations. How can it be impossible to collect them all together into a single Set?

We can draw these thoughts together very tidily as follows.

The paradox of the Set of all Sets: We both do, and do not, want to admit the existence of a Set of all Sets.

The affinities of this paradox with both Russell's paradox and the Burali-Forti paradox should be clear: with Russell's paradox, because the Set of all Sets would be, precisely, the Set of all Sets not belonging to themselves (no Set belongs to itself); and with the Burali-Forti paradox, because any attempt to collect together all the Sets would be, at the same time, an attempt to collect together all the ordinals indexing the stages of Set construction.[7]

It is sometimes said that we can escape the paradox by admitting the existence of a collection containing all the Sets but simply denying that it is itself a set of any kind. The term 'proper class' is often reserved for collections which are not sets. Cantor too had a term for such collections. He called them 'inconsistent totalities' (see above, Chapter 8, §5). But although distinguishing between collections of different kinds in this sort of way can have a perfectly legitimate rationale, especially in formal contexts, it is surely a mistake to think that it enables us to escape this particular paradox. There simply is no provision for distinguishing between those collections which are sets and those which are not. A set is supposed to be *any* collection of things. In any case, even if we did acknowledge the existence of proper classes, what could prevent us from building up a hierarchy of *them* and encountering an analogous paradox? (Thus what about the Proper Class of all Proper Classes?) We do best, surely, to try to ride the paradox and, instead of talking about the collection of all Sets as if it were something other than a Set, refrain from talking about it at all. There is no such thing.[8]

2 Ordinals as sets

Let us now return to the ordinals. The identity of each ordinal is determined, at the most fundamental level, by its predecessors. The most fundamental characterization of ω, for example, is this: ω is the ordinal whose predecessors are the natural numbers. To know the set of ordinals which precede a given ordinal is, in effect, to know *it*. Given the special concern which mathematicians have with structure, it is not surprising, then, that the suggestion should have been made, first by von Neumann,[9] that we should *identify* each ordinal with the set of its predecessors; and it is now standard practice to do this. We shall do likewise. Thus, for example, ω simply *is* the set of natural numbers {0, 1, 2, ...}, or in other words ℕ. Its successor, ω + 1, is {0, 1, 2, ..., ω}. And so forth. But what about the natural numbers themselves? Well, there is no reason why the proposal should not be extended to them, so long as we do not think that we are saying what the natural numbers 'really are' in some deep, philosophical sense. The proposal is a piece of mathematical legislation, to be assessed, if at all, in terms of its power, elegance, and beauty. That 5 = {0, 1, 2, 3, 4} is not a claim we would intuitively have wanted to make before: but nor, in mathematical terms, would anything have hung on our denying it.[10] What is interesting is what emerges when this proposal is traced down to base. 0 obviously has no predecessors. It must therefore be identified with the empty set, Ø. Thus

$$
\begin{aligned}
0 &&&= \emptyset; \\
1 &= \{0\} &&= \{\emptyset\}; \\
2 &= \{0, 1\} &&= \{\emptyset, \{\emptyset\}\}; \\
3 &= \{0, 1, 2\} &&= \{\emptyset, \{\emptyset\}, \{\emptyset, \{\emptyset\}\}\};
\end{aligned}
$$

$$
\omega = \{0, 1, 2, \ldots\} = \{\emptyset, \{\emptyset\}, \{\emptyset, \{\emptyset\}\}, \ldots\};
$$

and so on.

It follows that the ordinals are not only sets, but Sets. They are there in the 'V'-shaped hierarchy, one for each stage of Set construction. (We can think of them as running up the middle of the hierarchy. We might also, of course, think of them as marking the time-axis that I mentioned in §1.) I shall exploit this fact several times in what follows.

The Infinite

3 Cardinals. Measuring infinite sets

We saw in Chapter 8 that some infinite sets are bigger than others. Indeed there is no limit to how big an infinite set can be. There is, however, a limit to how small an infinite set can be. Any infinite set is at least as big as $\mathbb{N}$. This is an apt point at which to introduce some terminology. Any set that is either finite or the same size as $\mathbb{N}$ (that is, as small as its infinitude allows it to be) is said to be *countable*. Any set that is bigger than this is said to be *uncountable*.

Ideally, we should like some way of measuring infinite sizes more precisely, so that we can state exactly how big a given infinite set is, express comparisons of size, and suchlike. It was another of Cantor's major contributions to transfinite mathematics that he provided us with the wherewithal to do just this. He devised *cardinals* (or *cardinal numbers*). Precisely what these do is to enable us to measure how big sets are, including infinite sets, just as the natural numbers enable us to measure how big finite sets are, and the ordinals enable us to measure how long well-orderings are. Given any size that a set might be, there is a cardinal which acts as a measure of, or registers, that size. One says how big a set is by specifying the relevant cardinal. In the case of a finite set this will simply be the appropriate natural number; thus the cardinal which registers the size of the set of planets in the solar system is the number nine. (It follows that the natural numbers, as well as being included among the ordinals, are included among the cardinals.) The cardinals that register the sizes of infinite sets are called infinite cardinals. One cardinal is said to be smaller than another if it registers the sizes of smaller sets. The smallest infinite cardinal (that which registers the size of $\mathbb{N}$) is referred to as $\aleph_0$.[11]

Cantor proved that the infinite cardinals get bigger in discrete steps. After $\aleph_0$ comes $\aleph_1$. Then there is $\aleph_2$. Then there is $\aleph_3$. And so on *ad infinitum*. But there is also a cardinal that is bigger than all of these. For the cardinals are endless, in just the same way that the stages of Set construction (and thus the ordinals) are endless. And, for exactly analogous reasons, we can use the ordinals to index them. Thus the first cardinal to succeed $\aleph_0, \aleph_1, \aleph_2, \ldots$ is $\aleph_\omega$. After $\aleph_\omega$ comes $\aleph_{\omega+1}$, then $\aleph_{\omega+2}$, and so forth. In general, for each ordinal α, there is a cardinal $\aleph_\alpha$.

But what *are* the cardinals?

Here it is natural to turn once again to the ordinals. Those shown in Figure 8.5 (see above, Chapter 8, §4) are all countable. That is, none of them has more than $\aleph_0$ members. Eventually, however, there must be an ordinal which is uncountable – namely, the first to succeed all those that are countable (or, in other words, the Set of all those that are countable). It has $\aleph_1$ members. Later still, there is the first ordinal with $\aleph_2$ members. More generally, given any cardinal κ, there must eventually be an ordinal, and thus a first such, with κ members; this is part and parcel of what it is for the supply of ordinals to be never-ending. A natural proposal then, which has already effectively been adopted at the finite level, is once again to prescind from what would otherwise be mathematically irrelevant non-structural differences and to *identify* each cardinal with that corresponding ordinal. Let us proceed in this way. Thus the sequence of transfinite ordinals has, scattered along it, cardinals. The cardinals are those ordinals which are the first of their respective sizes. Between any two infinite cardinals there are many ordinals which are *not* cardinals. For example, between $\aleph_0$ and $\aleph_1$ there are such ordinals as $\omega + 1$, $\omega + 2$, $\omega \times 2$, ω^2, ω^ω and ε_0 (ε_0 is the first ordinal to succeed all of ω, ω^ω, ω^{ω^ω}, ...). $\aleph_0$ itself, of course, is just ω, the first infinite ordinal. We can now see that '$\aleph_0$', 'ω' and '$\mathbb{N}$' are three names for one and the same entity, to wit the Set of natural numbers.

We can also see that, given any infinite cardinal, there are two ordinals with which it is intimately associated. First, obviously, there is the ordinal which it actually *is*. But there is also the ordinal which is used to index it, or, if you like, to register how far it is along the sequence of infinite cardinals. In the case of $\aleph_0$ these two ordinals are, respectively, ω (for $\aleph_0$ *is* ω) and 0 (for its index is 0). One might think that the first of these ordinals must always come much later in the sequence of ordinals than the second. For how can the ordinal which a cardinal *is* not far outstrip the ordinal which registers how soon we encounter it? (There are, after all, many ordinals between the cardinals.) Nevertheless, mind-bogglingly, a cardinal κ exists that is sufficiently large for the two ordinals to co-incide and for it to act as its own index: that is, $\kappa = \aleph_\kappa$. It is as if this is a number *so* big that you need something that big to say how big it is! (In fact, there are infinitely many cardinals of this kind.[12])

Cantor went on to present a kind of arithmetic of infinite cardinals. He explored what happens when one infinite cardinal is added to another, when it is multiplied by it, when it is raised to the power of it, and so forth. Of course, given the standard definitions of these arithmetical operations, it does not make sense to apply them to infinite cardinals. But there are very natural ways of extending the definitions.

Addition and multiplication turn out to be rather unexciting. One might have expected, for example, that $\aleph_7 + \aleph_5 = \aleph_{12}$, or that $\aleph_7 \times \aleph_5 = \aleph_{35}$.[13] In fact, however, if two infinite cardinals are added/multiplied, the larger 'swallows up' the smaller and is itself the sum/product. Thus

$$\aleph_7 + \aleph_5 = \aleph_7 \times \aleph_5 = \aleph_7.$$

More generally:

if κ and λ are cardinals, at least one of which is infinite, and $\kappa \geqslant \lambda$,
then $\kappa + \lambda = \lambda + \kappa = \kappa \times \lambda = \lambda \times \kappa = \kappa$.

Exponentiation is much more interesting. Perhaps the most important result is the following:

$$\text{if } \kappa \text{ is an infinite cardinal, then } \kappa < 2^{\kappa}.$$

Raising 2 to the power of an infinite cardinal always issues in a larger cardinal. In particular, $\aleph_0 < 2^{\aleph_0}$. What is significant about this is that it is possible to prove that both the power-set of $\mathbb{N}$ and $\mathbb{R}$ have $2^{\aleph_0}$ members. The fact that $\aleph_0 < 2^{\aleph_0}$ is therefore a reflection of the fact that $\mathbb{N}$ is smaller than its own power-set and smaller than $\mathbb{R}$.

But how much smaller? Does $2^{\aleph_0}$ equal $\aleph_1$? $\aleph_{243}$? $\aleph_{821}$? $\aleph_{\omega^2}$? This is none other than Cantor's unanswered question (see above, Chapter 8, §3), the question that Hilbert took himself to have settled (see above, Chapter 9, §2). To say that the power-set of $\mathbb{N}$, or $\mathbb{R}$, is the next infinite size up from $\mathbb{N}$ is, in effect, to say that $2^{\aleph_0} = \aleph_1$.[14] This was the hypothesis put forward by Cantor, though he never succeeded either in proving it or in disproving it. It is known as the continuum hypothesis.

Cantor's continuum hypothesis: $2^{\aleph_0} = \aleph_1$.

The hypothesis derives its name from the fact that the real numbers (the members of $\mathbb{R}$) are used to register points on a continuum. Let us now look at it in somewhat greater depth.

4 The continuum hypothesis

To this day the hypothesis remains neither proved nor disproved. Cantor's question is still unanswered. However, *just* to say this would be to tell only part of the story and it would be seriously misleading. The suggestion would be that mathematicians simply have not been assiduous enough; if only they applied a little more effort they might be able to settle the question. But remember, mathematical practice here is guided by the nine axioms of **ZF**, which are intended to constitute a precise and formal description of what sets (or more specifically, Sets) are like. It is on this basis that an answer to the question must initially be sought. Yet Gödel proved, in 1938, that it would be impossible, using these nine axioms, to show that the continuum hypothesis was false, while Cohen proved, in 1963, that it would be impossible, on the same basis, to show that it was true – unless, what we very much hope is *not* the case, **ZF** embodies some inconsistency that we are not aware of, which may mean that the whole idea of a set must finally be dismissed as incoherent, in which case the question whether the continuum hypothesis is true or not ceases to be an issue.[15] Gödel's earlier work, which was mentioned in Chapter 8, §6 and to which we shall be returning in Chapter 12, had already shown that this situation was on the cards: so long as **ZF** *was* a consistent theory, there were bound to be set-theoretical questions that it was powerless to settle. The question about the exact size of $2^{\aleph_0}$ turns out to be such a question.

But what follows?

It certainly does not follow that the continuum hypothesis has been shown to be neither true nor false, whatever *that* might mean. Nor does it follow that it has been shown to be neither provable nor disprovable. For nothing has been shown about what might be done beyond the confines of ZF with new and more powerful axioms. Thus one possible reaction to the situation, pretty much Gödel's,[16] would be to insist that the question has an answer, determined by the mathematical reality that has been informally described in this chapter, and to conclude that ZF is simply an incomplete description of that reality: to address the question, we must bolster ZF by introducing some new fundamental principle about what Sets are like – a tenth axiom. Much current research in set theory is in fact concerned with exploring new axioms and seeing whether any of them might be strong enough to settle the question. For the most part these axioms concern very large cardinals.[17] The problem is that none of them has the intuitive appeal of the original nine axioms. We are not sure which of them are true, or why. So, on this understanding of the situation, it is a very real possibility that we simply do not have the relevant insight into mathematical reality to be able to determine whether or not the continuum hypothesis is true. One can, of course, be optimistic – as was Cohen. He suggested that we may eventually come to see the hypothesis not just as false but as *obviously* false.[18] But it seems more likely that we are in territory where our intuitions are beginning to run dry.

This in itself suggests another possible reaction to the situation, quite different, namely to insist that ZF does capture the essence of the intuitive conception of Sets outlined in §1, but to conclude that that conception is not a fully precise one: it suffers from an indeterminacy rather like the indeterminacy from which our conception of baldness suffers. There are some men, with suitably thinning hair, such that there is really no fact of the matter whether they are bald or not. We cannot call upon any feature of our conception of baldness to settle the issue. They are, simply, borderline cases. Similarly, according to this view, we cannot call upon any feature of our conception of a Set to settle whether $2^{\aleph_0} = \aleph_1$ or not. Of course, imprecision of this kind can always be artificially eradicated. We could, for example, simply stipulate that a man with such and such a number of hairs is to count as bald. Likewise, we might stipulate that $2^{\aleph_0} = \aleph_1$, or $\aleph_{243}$, or whatever (though some stipulations are ruled out; for example, the nine axioms of ZF preclude $2^{\aleph_0}$'s being $\aleph_\omega$). But the most significant thing about this would be precisely the fact that it was a stipulation.

There are other possible reactions to the situation. Their assessment cuts deep into the philosophy of mathematics, and into the philosophy of language, and indeed into set theory itself. But I shall not pursue them here.[19]

5 Further thoughts on the infinite by addition and the infinite by division

One source of interest in the continuum hypothesis, it might be suggested, is the way in which it bears on the fundamental distinction between the infinite by addition and the infinite by division. For, the suggestion might run, the hypothesis has to do with the exact relationship in size between $\aleph_0$ and $2^{\aleph_0}$: but $\aleph_0$ is the cardinal that measures the size of $\mathbb{N}$, and this – if we think in terms of an endless progression – embodies the essence of the infinite by addition; while $2^{\aleph_0}$ is the cardinal that measures the size of $\mathbb{R}$ and hence of any segment of $\mathbb{R}$, and this – if we think in terms of the points on a line – embodies the essence of the infinite by division. We can then give this a deliberately paradoxical twist by saying that the infinity of the infinitely small turns out to be bigger than the infinity of the infinitely big.

In fact, however, this suggestion is facile. The infinite by addition and the infinite by division are too loosely defined for any relationship between $\aleph_0$ and $2^{\aleph_0}$ to have such a direct bearing on them. For example, the basic idea underlying our intuitive concept of the infinite by division, as applied to a line for example, is simply that there should be, between any two points on the line, a third. As I pointed out in Chapter 4, §2, the precise counterpart of this idea is denseness, not continuity (in its technical sense). And for a mathematical embodiment of denseness it suffices to turn to the rational numbers; there is no need to invoke the reals. Yet, as the paradox of the pairs shows (see above, introduction, §1), there are no more rational numbers than natural numbers. It is not clear, then, why simple reflection on the concept of infinitude by division need take us beyond the countability of $\aleph_0$.

Conversely, the intuitive concept of infinitude by addition might *well* take us beyond that point, as soon as we take into account the ordinals. For the ordinals show us a kind of infinitude by addition that extends way beyond the natural numbers. This ties in with one of the intuitions that motivate the paradoxes of the one and the many, namely that what is truly infinite by addition must outstrip anything with a determinate size. It must not be in any way compressible into a set. A genuinely infinite totality is a many that is too big to be regarded as a one.

Of course, once we give fairly free rein to our intuitions in this way, allowing them at the same time to be enriched by mathematical knowledge, then our intuitive understanding of infinitude by division is itself liable to take another turn. Certainly continuity (in its technical sense) will seem to provide us with a richer and fuller description of those things that we ordinarily take to be infinite by division than mere denseness will. But just as our intuitive understanding of infinitude by addition propelled us beyond any straightforward mathematical characterization of it, so too our intuitive understanding of infinitude by division is liable to do the same. For how can a line, say, or the passage from A to B, be fully represented as

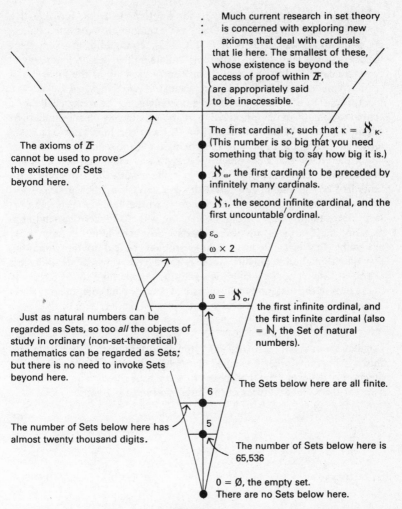

Much current research in set theory is concerned with exploring new axioms that deal with cardinals that lie here. The smallest of these, whose existence is beyond the access of proof within ZF, are appropriately said to be inaccessible.

The axioms of ZF cannot be used to prove the existence of Sets beyond here.

The first cardinal κ, such that $\kappa = \aleph_\kappa$. (This number is so big that you need something that big to say how big it is.)

$\aleph_\omega$, the first cardinal to be preceded by infinitely many cardinals.

$\aleph_1$, the second infinite cardinal, and the first uncountable ordinal.

ε_0

$\omega \times 2$

$\omega = \aleph_0$, the first infinite ordinal, and the first infinite cardinal (also $= \mathbb{N}$, the Set of natural numbers).

Just as natural numbers can be regarded as Sets, so too *all* the objects of study in ordinary (non-set-theoretical) mathematics can be regarded as Sets; but there is no need to invoke Sets beyond here.

The Sets below here are all finite.

6

The number of Sets below here has almost twenty thousand digits.

5

The number of Sets below here is 65,536

$0 = \emptyset$, the empty set. There are no Sets below here.

Figure 10.1

as a mere set of points? Ought we not to apply to the infinitely small the same intuition that was applied to the infinitely big, and say that a genuinely infinite totality, in this case a totality of points, is a many too big to be regarded as a one? There is a technical notion of *absolute continuity* which is pertinent here. It was developed by the mathematician Haussdorf, and is based on the absolute infinity of the ordinals. It incorporates precisely this idea, that there are more points on a line than can be gathered into any set.[20]

157

This idea is related to, though not inseparable from, an idea that surfaced many times in Part One, for example in Aristotle, Peirce, Bergson, and Gödel, namely that a line is something prior to – something that exists over and above – any set of points on it, a lesson that many of these thinkers took to be implicit in Zeno's paradox of the arrow. This seems to me to be correct. Points are where lines *do* things, such as stop, or meet. If one took the oppposite view, one might, for example, challenge a claim that I made in the Introduction, namely that a line must be infinite by addition if, beyond any two points on it, there is a third. For one might envisage taking a finite line and removing one of its end points; and one might then insist that what one was left with was a line that was still finite but such that, beyond any two points on it, there was a third. But how can a finite line be missing an end point? Its end point is neither more nor less than where it stops. Points are, precisely, where lines do such things as stop. Lines themselves, for that matter, are where surfaces do such things as stop; and surfaces are where bodies do such things as stop.[21] By extending this principle we might eventually be led to the somewhat Hegelian thought that only the whole is non-derivatively real; anything less is only an aspect of the whole, 'where it does something'. We are now in the realms of the metaphysically infinite of course. And consequently these are issues to be reserved for later (see below, Chapter 15, §3).

I shall draw this chapter to a close by drawing attention to Figure 10.1. This shows a diagram of the 'V'-shaped hierarchy of Sets (modelled on a similar diagram of Rucker's[22]). It is meant to serve as a convenient reminder of some of the salient features of the discussion in this chapter.

CHAPTER 11

The Löwenheim-Skolem Theorem[1]

What is that thing which does not give itself, and which if it were to give itself would not exist? It is the infinite! (Leonardo da Vinci)

We say: but that *isn't* how it is! – it *is* like that though! and all we can do is keep repeating these antitheses. (Ludwig Wittgenstein)

1 An introduction to the Löwenheim-Skolem theorem. Reactions and counter-reactions

At the end of Chapter 8, I mentioned two results in mathematical logic, those established by Skolem and Gödel, which both have a direct bearing on the infinite. Each is a rich fund of material lying ready to be woven into our understanding of the infinite. Each can be used to strengthen our grasp of the basic issues and problems that have begun to arise. I shall devote this chapter to a study of Skolem's result, the next to Gödel's.

Before I begin, I should emphasize that the two results are as far beyond controversy as any piece of pure mathematics can be. It is true that I shall be taking for granted certain methods of proof that have been challenged (for example, by intuitionists). But for current purposes we do best to take the results as a kind of datum. We can think of the really interesting philosophical dialectic as beginning at the point where their import is being probed and they are being used to illustrate, defend, or challenge non-mathematical ideas.

I have referred in various different ways to what it was that Skolem proved, but since in fact he was embellishing a result that had earlier been established by the mathematician Löwenheim, his theorem is usually referred to as the Löwenheim-Skolem theorem. It is beyond the scope of this book to go into the details of the theorem, but I shall try to present its essence.[2]

Suppose the iterative conception of a set which was outlined in the last chapter either to be, or somehow to have been made, fully determinate. In particular suppose that every sentence in the language in which **ZF** is

159

couched (including the sentence stating the continuum hypothesis) can now unproblematically be regarded as a true or false statement about what Sets are like. Now there are, in the primitive vocabulary of this language – I shall explain shortly what I mean by 'primitive' – only two words that give away its subject matter, namely 'Set' and 'member'. It otherwise consists of elementary words like 'not', 'and', 'every', and 'is', which can be used to talk about anything whatsoever. This claim may sound outrageous when one considers sentences like

$$2^{\aleph_0} > \aleph_0,$$

or

Between any two cardinals there are infinitely many ordinals,

or even

$\{\emptyset\}$ has only one member.

But this is why I talk of the language's 'primitive vocabulary'. I am alluding to the fact that all the other peculiarly set-theoretical terminology can ultimately be defined in terms of 'Set' and 'member' and is, in principle, dispensable. Those two words constitute a kind of lexical rock bottom. Thus instead of '$\{\emptyset\}$' we could write 'the Set whose only member is the Set which has no members'. Similarly with 'ordinal', '$\aleph_0$' '$2^{\aleph_0}$', and the rest, though in their cases the resulting expressions would be horrendously complex.

Suppose now that you had no idea what either 'Set' or 'member' meant, beyond knowing that they were mathematical terms. Think of them as two utterly alien terms, say 'zad' and 'nanpal'. How much could you determine about their meaning from being presented with true sentences from this language?

You might be told, first,

No two zads have exactly the same nanpals,

then

There is one zad which has no nanpals.

Already your picture of what is being described would be that little bit fuller and that little bit clearer. As you were presented with more and more truths, including perhaps all the axioms of **ZF**, then, bit by bit, your picture would become fuller and clearer still. But how full? How clear? Would you ever reach the point of knowing for sure what the intended interpretation of 'zad' and 'nanpal' was?

No. This is where the Löwenheim-Skolem theorem comes in. It entails that, however many true sentences from this language you were presented with (even if you were somehow presented with them all), you could never rule out the hypothesis that the things being described (the 'zads') were natural numbers.

Reconsider the two sentences above. They do nothing to disabuse such a hypothesis. 'Nanpal' could mean 'predecessor' for example. It is true that a set-theoretical sentence would eventually come to light to rule out *this* particular interpretation. The following is an example:

There is more than one zad which has only one nanpal.

(This is true under the intended interpretation, but false when 'zad' means 'natural number' and 'nanpal' means 'predecessor'.) But, according to the Löwenheim-Skolem theorem, the conjecture that there is *some* such interpretation of 'nanpal' could never be ruled out.

Admittedly the natural numbers are themselves Sets (or have been construed as such (see above, Chapter 10, §2)). But they are obviously not all of them. More pertinently, there are only countably many natural numbers. They constitute an almost negligible drop in the Set-hierarchical ocean. They are just a small initial segment of the endless backbone of ordinals. Yet no amount of set theory, it transpires, can force us to recognize that its subject matter comprises any more than that, let alone that it is the full panoply of Sets with all their complex interrelations.

Now it is not immediately obvious what the significance of this is, or indeed whether it is particularly significant at all. Imagine an interlocutor coming in at this point and responding as follows:

Obviously the truths in the language of **ZF** are truths about the full hierarchy of Sets partly because of what 'Set' and 'member' *mean*. It is hardly surprising that when we prescind from that meaning and enquire into the structure of what is left, that structure places quite meagre constraints on what the subject matter of the language can be said to be. All the Löwenheim-Skolem theorem has done is to show something about just how meagre. It has revealed a technical sense in which there is more to Sets than what can be truly said about them in this language – and likewise more to the meaning of 'Set' and 'member' than how they figure in such truths.

In the end I think this response is the right one. But as presented, it is too cavalier. The Löwenheim-Skolem theorem carries a very real threat, and it is important to see why.

It is all very well appealing uncritically to the meaning of 'Set' and 'member'. But precisely one of the distinguishing features of mathematical expressions seems to be that their meaning has to be grasped in terms of how they figure in the truths of a formal theory. The word 'Set' is not like the word 'apple' for example. If you did not know what the word 'apple' meant, then someone could give you an indication by picking up a specimen and pointing to it. But they could not pick up a Set and point to it. To give you an indication of what 'Set' meant they would basically need to say a whole lot about Sets – the kind of thing we were just envisaging (and indeed the kind of thing I attempted in Chapter 10). Thus, if the question arose as to how much you could learn about Sets by being presented with true sentences from the language of **ZF**, it would be reasonable to hope that, in principle, and in the end, you could learn everything. What, after all, would a full account of what Sets are like be but some kind of compilation of such sentences? Yet the Löwenheim-Skolem theorem seems to dash the hope, and leaves us with the problem of accounting for the meaning of 'Set' and 'member' in some other way. The worry, which was not addressed in the interlocutor's response above, is that there seems to be no other way of accounting for their meaning that is not utterly mysterious – for example, that their meaning is something to which we gain access by means of some kind of gnostic union with mathematical reality.

The interlocutor might come back at this point with the following response:

There is no special mystery in how we come to understand what 'Set' and 'member' mean. *Partly* it is a matter of subjection to some formal mathematical theory. But partly it is a matter of less precise, though no less important, analogies, comparisons, hints, suggestions. We might be asked to think about throwing a lasso round a bunch of objects, or putting them in a box. We might be asked to think about joining a club. Granted some minimal understanding of what kind of thing a set is, the truths in the language of **ZF** can then be used to determine precisely what Sets are like, right the way up through the hierarchy.

At this point, however, the Löwenheim-Skolem theorem strikes another blow. It has a more powerful consequence than the one outlined above; and this consequence seems not only to undercut what has just been said but to raise a whole complex of new paradoxes and conundrums of its own. It is this. There is a Set *M* that satisfies the following two conditions.

(i) *M*, though infinite, is only countable. In fact it is a slice off the very bottom of the 'V'-shaped hierarchy, insignificantly small in comparison with what is left behind (see Figure 11.1).

(ii) If 'Set' and 'member' had precisely their intended interpretation, save that they were restricted in their application to the members of *M*, then all the truths in the language of **ZF** would still be true.

It is as if, when we turn to the hierarchy of Sets, we are forced to conclude that all but a very tiny, nay infinitesimal, portion of it, snuggled in the vertex, is dead wood: it would be all the same if nothing other than this infinitesimal portion existed.

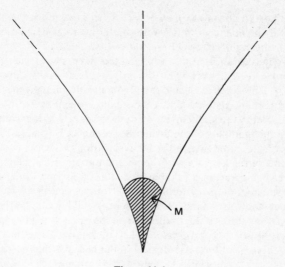

Figure 11.1

Certainly this seems to put paid to the thought that a minimal under-standing of what kind of thing a set is would be enough to guarantee that the truths in the language of **ZF** should be interpreted as truths about the full hierarchy of Sets. We need some more radical way of addressing the threat about meaning and understanding that the Löwenheim-Skolem theorem poses. I shall return to this task later on in the chapter.

First, however, we must address the new and even more alarming aspect of the theorem that has now come to light. It looks, to put it bluntly, absurd. How can all Sets bar the members of M be irrelevant to the truths in the language of **ZF**, given that some of those truths involve explicit reference to portions of the hierarchy beyond M? Think about statements concerning larger and larger cardinals. Or think about the statement that some Sets, such as the power-set of ω and $\aleph_1$, are uncountable. How could statements such as these be true if there were only countably many Sets altogether? The apparent contradiction here is often referred to as Skolem's paradox, Skolem himself being the first to draw attention to it.[3] It can in fact be solved in a rigorous and uncontroversial way, and it is important, before we proceed any further, to see how.[4]

2 The solution to Skolem's paradox. Scepticism and relativism

Let us focus on a particular sentence that helps to fuel Skolem's paradox, for example:

The power-set of ω is uncountable.

What we have to do is to explain how sentences such as this could remain true even if the language of **ZF** were reinterpreted as dealing exclusively with the members of *M*, which we will call '*M*-Sets'.

Now we must not forget that what matters here is what such a sentence would look like if it were spelt out in gory detail using only primitive vocabulary. This sentence would say (unintelligibly enough) that there was a Set of a certain complex kind – a Set whose members were those Sets whose members were all members of another Set of a certain complex kind – and that the members of the former Set could not be paired off with the members of the latter, where the idea of pairing off would itself be spelt out in terms of the availability of some appropriate Set of pairs, the pairs themselves being Sets of a suitable kind. Very well: now, what would the sentence say under the reinterpretation? Not just that there were two *M*-Sets of such and such a kind whose members could not be paired off. The idea of pairing off would itself have to be spelt out in terms of the availablity of some appropriate *M*-Set. Once we think this through, however, there is nothing paradoxical in the fact that the sentence should still be true. It is as if, *within M*, there is something which looks like ω and there is something which looks like its power-set and it looks as though their members cannot be paired off. When we step outside *M*, or perhaps we should say, from our vantage point already outside *M*, we can see that what looks like ω down there really is ω, whereas what looks like its power-set is a Set that is in fact only countable, and the reason they look to be of different sizes is that there is no Set of pairs to establish this countability *within M*. There *is* such a Set of pairs, but it is not an *M*-Set.

This dispels the paradox from a technical point of view. However, it creates fresh worries of its own. Consider: if somebody suffered from the illusion that the subject matter of **ZF** really *was M*, then they would make exactly the same set-theoretical claims as we do. They would describe *M* in exactly the same way that we describe the full hierarchy. They would talk about 'endless stages of Set construction', 'the backbone of ordinals', 'uncountably big cardinals', and so forth, and their illusion might never show up. They would even be able to distinguish their own '*M*', at the bottom of what they understood to be the full hierarchy of Sets, entertaining the same sceptical doubts about it as we are now entertaining about the real *M*. But these sceptical doubts can be pushed even further. Not only can we acknowledge the possibility that somebody else is under such an illusion, maybe *we* are. This is somewhat reminiscent of the horrific possibility mooted in a number of recent philosophical discussions: that a human brain might be kept alive in a vat and be so manipulated by scientists as to give the subject the illusion of living a perfectly normal life with a perfectly normal body. The sceptical challenge comes in the question: how do I know that *I* am not in that position? Similarly: how do we know that what *we* understand to be the full hierarchy of Sets is not really some countably big (and thus infinitesimal) portion of all the Sets there are? How do we know that what *we* take to be uncountable may not appear countable from elsewhere?

In fact, if we took seriously the temporally charged metaphor of Set construction, then we might want to go further. We might want to say that we not only do not know that we are not in that position, we do know that we are. For if the hierarchy of Sets is never given in its entirety, but is in a continual state of construction, then we can be sure that whatever Sets have now been constructed, there are more to come; and these may well provide us with the wherewithal to represent as countable some Set that we now take to be uncountable. We must in any case concede a kind of relativism, it seems. For has it not been shown that a Set is neither countable nor uncountable, full stop, but only from this or that point of view, or relative to this or that collection of Sets?

I do not believe that any such thing has been shown. Nor indeed do I believe that such relativism is coherent (though it is interesting to note that Skolem himself espoused a version of relativism when first drawing attention to these issues[5]). Nor do I believe that taking the temporal metaphor this literally is coherent. Nor do I believe that there is any coherent formulation of the original sceptical doubts. I take all of these to rest on a common misconception, as I shall now try to explain.

3 Scepticism and relativism rebutted

It is certainly true that we can recognize relativized notions of countability and uncountability. A Set may be uncountable relative to M, for example, but countable relative to some more inclusive Set N. However, relativism goes further and says that we cannot recognize any *un*relativized notions of countability and uncountability. The suggestion is that when we take a particular Set to be uncountable, full stop, we are deluding ourselves: it is really only uncountable relative to what happens to be our point of view.

What is incoherent both about this relativism, and about an overly literal construal of the temporality of Set construction, and about the original sceptical doubts, is that all three trade on a notion of 'our point of view' which we simply have no way of making intelligible to ourselves. Our point of view is supposed to do two things: (i) it is supposed to constrain what we (now) understand to be all the Sets there are (or in other words, what we (now) understand by the term 'Set'); and (ii) it is supposed to be something that we can at the same time recognize, or at least think of, as so constraining us. But (i) already precludes (ii). For if our understanding really cannot reach beyond certain narrow confines (if only for the time being), then, in particular, it cannot reach to the fact that it cannot reach beyond those confines. We simply cannot make sense of our being subject to limitations in this way. We cannot make sense of the idea that our understanding should not be able to embrace *all* Sets. This is related to the fact that, when we reflect on what we do understand to be all Sets, we cannot see them as constituting a single (limited) collection – a further Set.

Of course, a possible retort here will be that, although *we* cannot see that such a Set exists, it is still there to be seen from a higher vantage point, or, if it is not there yet, it will eventually be constructed. (It is not as if the idea of overseeing the construction, within time, of new and bigger Sets is particularly fanciful. Is this not a good way to view what happens when we come to accept, as a new axiom, a sentence that postulates such Sets?)

However, it is self-stultifying for us to try to accommodate this retort. (It is self-stultifying to say, and to mean at all literally, that not all Sets have been constructed at any given point in time. This is itself a generalization that we are *now* making about *all* Sets.) The point is this. What *we* understand to be all the Sets there are we understand to be *all* the Sets there are.

This might not seem enough to counteract the original sceptical doubt that we might yet be wrong: what we understand to be all the Sets there are might not *in fact* be. (We might be like the person who believes that 'Set' is used to refer to *M*-Sets.) But how are we to entertain this doubt? Where, so to speak, are we to entertain it? If the reply is, 'From our limited point of view,' then we cannot retain any grip on this supposedly deep notion of what is 'in fact' the case. But if the reply is not that – if, in other words, we are not subject to such a limitation – then the doubt has already lost its rationale. The sceptic's hope is a forlorn one: to urge us to see ourselves as having a limited point of view which we cannot rise above – but which we cannot otherwise see ourselves as having.

Of course, I do not deny that the sceptic seems to have got hold of something. The doubts continue to gnaw. There is still an urge to say, 'But even so, we *might* have a limited point of view.' Our problem, however, is in giving further voice to this doubt. It is as if we are trying to say something that cannot quite be said. It is as if there is a real insight here that cannot quite be articulated or entertained. These are suggestions that I shall later be taking very seriously (see below, Chapter 13).

But meanwhile we do best just to turn our backs on this kind of scepticism, and on the relativism, and to insist that we use the word 'Set' quite simply to talk about *Sets* – all of them. (And we can show that the power-set of ω is uncountable, full stop, not just uncountable relative to this or that.) The fact that there is a countable Set *M* which acts as a miniature model of the entire hierarchy of Sets is of purely technical interest. It ought not to affect our view of the set-theoretical statements that we wish to endorse, nor our view of what they mean.

In saying this, I am returning to the rather nonchalant attitude taken by the interlocutor in §1, first responding to the Löwenheim-Skolem theorem. This is therefore an apt point at which to come back to the worries about

166

the meaning of 'Set' and 'member', and about how we come to grasp that meaning, which were originally brought against this attitude and which were left waiting to be addressed.

4 Meaning and understanding. The Löwenheim-Skolem theorem finally defused

What we have to do, I submit, is to distinguish between two quite different senses in which the truths of set theory might be said to fix the meaning of 'Set' and 'member'. The sense in which we think they must do is different from the sense in which the Löwenheim-Skolem theorem establishes that they do not. Once we have made this distinction, the threat that the theorem posed will have been averted.

The first sense: The sense in which we think the truths of set theory must fix what 'Set' and 'member' mean is the sense in which, if you knew nothing about what Sets were like, and were trying to understand (or to learn how to use) the terms 'Set' and 'member' from scratch, you would, ultimately, have nothing to go on but these truths.

Comment: This was the point being made when the contrast between 'Set' and 'apple' was drawn in §1.

The truths of set theory do indeed fix what 'Set' and 'member' mean in this sense. Of course, all sorts of subtleties would be involved in your assimilating these truths, and I do not mean to suggest that you would have no alternative but to learn to repeat as many of them as possible, parrot-fashion. Analogies with lassos, clubs, and suchlike might help to ease you in, and indeed provide you with suitable motivation. But there would be no way of bypassing the truths to see directly what Sets were like.

The second sense: The sense in which the Löwenheim-Skolem theorem establishes that the truths of set theory do not fix what 'Set' and 'member' mean is a very special, technical sense that involves relations between the language of **ZF** and its various possible interpretations, when these are themselves being construed as elements of mathematical reality. There are unintended interpretations that make the same sentences come out true.

It is curiously tempting to assimilate these two senses, which helps to explain our original sense of conflict and threat. It is tempting to think that learning what 'Set' and 'member' mean from scratch involves, precisely, having a prior grasp of these various candidate interpretations and then seeing which fits. But, on further reflection, it is obvious that this is a crazy view of what is involved. It is crazy because the interpretations them-selves, as elements of mathematical reality, can only be understood in set-theoretical terms – by someone who already has some mastery of set-theoretical language. This is especially transparent in the case of *M*. But it is also true of the interpretation involving the natural numbers, one of whose significant features is that the set (indeed Set) of natural numbers is countable. (Countability is a set-theoretical notion.) You cannot fully know what these interpretations are like unless you are already well immersed in some set-theoretical practice. As Putnam puts it, interpretations are not 'lost noumenal waifs', they are themselves 'constructions within our theory'.[6]

It is worth remarking parenthetically that it would, by the same token, be utterly disingenuous to make the following sceptical suggestion: that, because more than one of these interpretations fits, perhaps we do not really know what 'Set' and 'member' mean (or perhaps we understand them differently from one another). Even by mooting the possibility we belie it. This is just another aspect of the way in which scepticism founded on the Löwenheim-Skolem theorem is self-stultifying. The point is much as it was before. To toy with these sceptical possibilities is itself to engage in set theory and to think (straightforwardly) about Sets.

The two senses are quite distinct then. There are two corresponding senses in which you might be said to know what 'Set' and 'member' mean. You might be said to know what they mean in the sense that you know how to use them properly, and in particular you know which set-theoretical sentences you are entitled to assert (in other words, which are true); this is something that obviously admits of degrees. Or you might be said to know what they mean in the sense that you are sufficiently mathematically sophisticated to be able to grasp the differences between their various possible interpretations and you know which is the intended one. These are not the same, for there is no reason to suppose that knowing how to use the terms involves matching them with, or thinking consciously about, any interpretation.

Once we have drawn these distinctions, we have a way of understanding what is involved when you first master set theory and come to know what 'Set' and 'member' mean which takes it right outside the province of the Löwenheim-Skolem theorem. There is no problem in saying that this is something you can do only by being subjected to the truths of set theory. This is to be taken in a very low-key and mundane way. The idea is simply that you have to be initiated into a mathematical practice by being shown it in operation. You observe what is done and learn to do the same. (This is the kind of thing that Wittgenstein might have said (see above, Chapter 9, §3).) However exactly the whole thing works, there is nothing in the Löwenheim-Skolem theorem that makes it especially problematical.

After all, similar considerations might be brought to bear on all the other words in the language of **ZF**, the words like 'and' and 'all' that are not peculiarly linked with its subject matter. They too need to be mastered and understood. I have been taking for granted throughout this chapter this demarcation between 'Set' and 'member' on the one hand and the rest of the language's primitive vocabulary on the other. (I have implicitly taken it for granted elsewhere too. In saying that **ZF** has nine axioms, for example, I mean that it has nine axioms governing the terms 'Set' and 'member'. Fundamental principles governing the rest of its vocabulary are taken for granted.) The demarcation is just an application of a more general distinction that is customarily drawn in mathematical logic and philosophy, the distinction between what is logical and what is extralogical. But although we have intuitions about this distinction, it has proved notoriously difficult to provide an uncontroversial account of how exactly to draw it. Certainly the loose characterization of what makes a term logical that I mentioned in §1 (that it 'can be used to talk about anything whatsoever') does not take us very far. It may convince us that 'Set' is extralogical. But what about 'set'? (It was an integral part of Frege's project to regard the concept of a set as a logical one (see above, Chapter 8, §2).) Then again, if 'set' is logical, ought not 'Set' to be logical too, the latter being, in some sense, a pure and refined version of the former? Correlative with these problems is the fact that no very satisfactory and full account exists of what the philosophical significance of the logical/extralogical distinction is, or at least not one that meets with universal acclaim. There is therefore considerable room for doubt about whether much of *philosophical* interest can hang on it. And this in turn helps to take even more sting out of the Löwenheim-Skolem theorem. (Why should it be important what happens when the interpretation of just *these* terms is up for grabs?)

To sum up: none of the threats that the Löwenheim-Skolem theorem posed has proved serious. The original unflustered response to the theorem made by the interlocutor in §1 has been vindicated. There is, in the relevant *technical* sense, more to the meaning of 'Set' and 'member' than shows up in the truths of set theory. But this, like the very content of the Löwenheim-Skolem theorem itself, so far from making us doubt our own understanding of what Sets are like, is something that we can only appreciate in terms of that understanding. (It involves us in actually exercising such concepts as that of countability.)

Our understanding of what Sets are like is part and parcel of our facility with handling the language of set theory – our being able to use the terms 'Set' and 'member' correctly. We have learnt to use 'Set' when referring to Sets and 'member' when referring to their members. Any hint of triviality here is intended. The point is simply not to let the Löwenheim-Skolem theorem, or anything like it, undermine a certain mathematical self-confidence.

5 A lingering paradox

The Löwenheim-Skolem theorem has, I hope, been defused, in the sense that any special threats that it posed have been averted. To that extent, we can indeed afford to be self-confident about our set-theoretical practice; *to that extent*. The problem is that self-confidence and self-consciousness make notoriously bad bedfellows. And there is a kind of self-consciousness that is supposed to be integral to set theory. It has already proved unsettling. It gave rise to perhaps the purest of the paradoxes of the one and the many, namely the paradox of the Set of all Sets (see above, Chapter 10, §1). Moreover, the Löwenheim-Skolem theorem, precisely in forcing us to think about our own use of set-theoretical terms, as we have been doing, serves to heighten this self-consciousness. It may not bring any special problems of its own. But it does, I want to conclude, help to exacerbate some of our original and deepest perplexities concerning the infinite. At the end of the day, there is a genuine paradox in the air here.

The point is this: in framing an appropriate response to the Löwenheim-Skolem theorem, we are forced to think self-consciously about the intended interpretation of the language of ZF – how, for example, it differs from the interpretation involving *M*. It is all very well saying that we use 'Set' when referring to Sets and 'member' when referring to their members. But, quite apart from the fact that this seems trivial, there is a further crucial respect in which it does not do full justice to how we use the terms: it does not properly differentiate between us and anyone using 'Set' and 'member' to deal only with *M*-Sets. The important point is that we use them to deal with *all* the Sets, something that we naturally express, particularly when we have the contrast with *M* in mind, by saying that our subject matter is the full hierarchy. But what kind of thing is the full hierarchy? Is it a thing at all? Presumably not, if there is no Set of all Sets. For a hierarchy can have no more definite an existence than the collection of things within it. But then with what right do we refer to the full hierarchy (as I have done several times throughout the last two chapters)? Surely not with any, if what we say is to be taken at face value. But how else is it to be taken? What is going on in these constant allusions? What are we doing when we – so we think – contrast the full hierarchy with small portions of it? Once again, it seems that we may be trying to express the inexpressible.

The discussion here is indeed intimately related to the earlier discussion of scepticism. For it is the same compelling urge to step back and reflect self-consciously on our own set-theoretical practice that makes us think both that there must be a definite grouping together of all that we are dealing with and, correlatively, that this grouping together might exclude what it could just as well have included (indeed *will* do so; it will exclude itself). The thought persists that we have a limited point of view, though there is no satisfactory way of saying so.

Of course, we could just hold back from referring to the full hierarchy, and at the same time resist the temptation to entertain such thoughts. Or we could allow ourselves to refer to it, and then attempt to explain how what we are saying is not, after all, to be taken at face value: it is a *façon de parler*, to be paraphrased in this or that way. (In line with the suggestion that I made in Chapter 10, we might even think of the concept of the full hierarchy as a Kantian Idea of reason.) But our self-consciousness will continue to whittle us. Our problem is, in a sense, *the* problem of the infinite. It rests on the fundamental paradox of the one and the many, that we both do and do not want to recognize unity in (truly) infinite multiplicity – or, more dramatically, that we seem both compelled to recognize unity there and compelled not to. We are also, of course, in the realm of the paradoxes of thought about the infinite. We seem forced to recognize the full hierarchy of Sets as something that we cannot – and *a fortiori* cannot be forced to – recognize as anything. Again, when we deny that there is a Set of all Sets, it seems fundamentally different from denying that there is a little green Martian behind the settee. We have a real sense of first getting into focus (seeing and acknowledging) what it is whose existence we are about to deny, then thinking, 'It is this, the totality of what we are talking about when we engage in set theory, our very subject matter conceived as a whole; *this* is what does not exist.' But this is absurd, in just the same way that it is absurd to grasp the (truly) infinite as that which is ungraspable. Yet this is what we seem to have done. We seem to have focused attention on the (truly) infinite as something that is not even there.

It will not help to resurrect some kind of actual/potential distinction, so that we can imagine ourselves glancing across time and focusing attention, as if *sub specie aeternitatis*, on what is simply not there *now*. For, however literally or metaphorically we want to take this, whether or not something is there now must be, in this context, a matter of whether or not we can now focus attention on it. It is a minimal sense of existence that is at stake. Even if the future is not (yet) the present, it is still there to be acknowledged – now – *as the future*. We can – now – think about time in its entirety (or else the proposal loses its force). There is no point in granting the hierarchy an existence even as potentially infinite, for that would still make it available as an object of mathematical study, just like the Sets that comprise it. Somehow we have to come to terms with the fact that there is, strictly speaking, no hierarchy. There is no Set of all Sets. Our best advice is surely that offered by Wittgenstein: to pass over in silence what we cannot talk about.[7]

CHAPTER 12

Gödel's Theorem

The human mind is incapable of formulating ... all its mathematical intuitions, ie., if it has succeeded in formulating some of them, this very fact yields new intuitive knowledge, eg., the consistency of this formalism. This fact may be called the 'incompletability' of mathematics.

(Kurt Gödel)

1 Introduction: the Euclidean paradigm

Gödel's theorem is one of the most profound results in pure mathematics. When it was first published, in 1931, it had a devastating impact. On the one hand, it laid waste a variety of firmly held convictions and initiated a struggle that has been going on ever since to come to terms with its mathematical and philosophical implications. On the other hand, it took the breath away for its sheer beauty. My aim in this chapter is to present an outline of the theorem and to spell out some of its implications for our own enquiry.

In a nutshell, it concerns the Euclidean paradigm – the paradigm of axiomatization. It is possible, we know, to devise a finite stock of fundamental principles or axioms from which all of the infinitely many truths of Greek geometry can be derived: this is the Euclidean paradigm.[1] Prior to 1931 it had been assumed by many that what was possible in geometry must be possible anywhere else in mathematics (and perhaps in non-mathematical contexts too); the paradigm must represent the very essence of mathematical method.[2] One of the reasons for this relates back to our discussion in the last chapter. Suppose we grant that the meaning of a mathematical expression has to be grasped in terms of how it figures in the truths of a formal theory. Then must there not be some way of 'capturing' these truths and providing them with a finite characterization – precisely what an axiomatization (and that alone?) can supply? How else could anyone assimilate the truths and grasp the expression's meaning? Again, relatedly, do we have any sense of mathematical truth apart from mathematical provability? When we say that a given mathematical statement is true, do we not mean (or at least imply) that there is a formal and precise proof of it? If so, it seems that there must be a finite specification of what the relevant canons of proof are – what the fundamental principles are to which we may ultimately appeal.

But Gödel's theorem belies these thoughts. What is possible in geometry is not possible in, for example, set theory. We have already seen that **ZF**, if it is consistent, is powerless to settle the truth or falsity of the continuum hypothesis. What Gödel's theorem shows is that there is an incompleteness here that is irremediable. Even if further axioms were added to **ZF**, and even if it became possible to settle that particular question, still, provided that no inconsistency had been introduced and provided that the axioms were still only finite in number,[3] there would be *some* true set-theoretical statements that could not be proved. (I am continuing to take for granted that every set-theoretical statement can unproblematically be regarded as true or false.) Set theory cannot be completely axiomatized.

It does not follow that set-theoretical truth is different from set-theoretical provability. Let us refer to a finite, consistent collection of axioms as an *axiomatic base* (or just *base*, for short). Then all that follows is that set-theoretical truth is different from provability using this or that particular axiomatic base. We must distinguish carefully between the following two claims:

(1) Given any axiomatic base *A* for set theory, there is some true set-theoretical statement *s*, such that *s* cannot be proved using *A*;

and

(2) There is some true set-theoretical statement *s* such that, given any axiomatic base *A* for set theory, *s* cannot be proved using *A*.

(1) is what Gödel's theorem establishes. It means that any axiomatic base has certain limitations. (2) would be a much stronger claim, to the effect that there was some true set-theoretical statement which was absolutely unprovable – which could not be proved using *any* base. We have no reason to suppose that (2) is true. We are still free to insist that a true set-theoretical satement must be provable in the sense that there must be a proof of it using some base or other. The point is simply this: no single base suffices for proving *all* these truths.

But that in itself is enough to put paid to a crude axiomatic view of mathematics. I mean the view that the truths comprising any given branch (geometry, analysis, set theory, . . .) are always those statements that can be proved using some particular axiomatic base. This is true of Euclidean geometry. But it is not true of set theory. Gödel's theorem also puts paid to some of the deeply entrenched intuitions aired above. It presents a real challenge.

Before I proceed to address this challenge, I shall try to present a rough sketch of why the theorem holds, to provide a focus for the discussion.

173

2 A sketch of the proof of Gödel's theorem[4]

I have so far been concentrating on the application of Gödel's theorem to set theory. But its most fundamental application is to *arithmetic*, by which I mean the theory of the natural numbers and the basic operations that apply to them, such as addition and multiplication. The theorem applies to set theory simply because set theory incorporates arithmetic (or, if you like, arithmetic can be reduced to set theory – the natural numbers can be identified with Sets (see above, Chapter 10, §2), and the relevant arithmetical operations can be defined in set-theoretical terms). I should acknowledge, in saying this, that I am taking for granted the same thing about arithmetic as I have been taking for granted about set theory: that each of its statements can unproblematically be regarded as true or false. (This assumption will come under scrutiny in §3.) It is arithmetic, first and foremost, that is shown not to be completely axiomatizable. How?

One thing that would show this would be a demonstration that there could not be a *decision-procedure* for arithmetical truth. By a decision-procedure for arithmetical truth, I mean a purely mechanical, step-by-step procedure for determining whether or not an arbitrary arithmetical statement was true or false, something demanding no mathematical insight or ingenuity, something guaranteed to produce an answer after a finite number of steps, something of the kind that a computer might, in principle, carry out. For if there *were* a complete axiomatization of arithmetic – an axiomatic base from which all arithmetical truths could be derived – then there would be such a decision-procedure. The procedure would be to search systematically through all the possible proofs that used the given axiomatic base, perhaps in order of increasing complexity, until either a proof that the statement was true or a proof that it was false came to light; one or other must eventually do so. (Of course, this procedure would be hopelessly unwieldy, even for a computer. It would be of purely theoretical interest. But that is not the point.) So, *could* there be a decision-procedure for arithmetical truth?

Certainly we are not in possession of one. If we were, we would no doubt have tried putting it to effect (unless it was, in fact, hopelessly unwieldy) to settle various outstanding unsolved arithmetical problems. (Goldbach's conjecture that every even number greater than 2 is the sum of two primes is an example that we encountered earlier. It is not known whether or not that is true.) Some of what we know in arithmetic, for example that $7 + 5 = 12$, we know by having effected a suitable limited decision-procedure. But other things that we know, for example that every natural number is the sum of four squares (in the way that, say, $21 = 16 + 4 + 1 + 0$), we know only because some mathematician, in this case Lagrange, was skilled or inspired enough to divine the proof. We have no general decision-procedure for arithmetical truth. But the question is not whether we do have. The question is whether we could possibly have. And if the answer is no, then Gödel's theorem is established.

To show that the answer is indeed no, we first consider those expressions in the language of arithmetic that pick out properties of the natural numbers, for example 'is odd', 'is a prime', 'is the sum of four squares' and '< 2'. Let us call these expressions 'number-predicates'. Each number-predicate can be said to define a set of natural numbers, namely the set of numbers of which it is true. The four number-predicates cited above define, respectively: the set of odd numbers; the set of primes; the set of *all* the natural numbers (for every natural number is the sum of four squares); and the set whose sole members are 0 and 1 (for only 0 and 1 are less than 2). But note: not every set of natural numbers can be defined by a number-predicate in this way. Some are too 'untidy'.

Now it is possible to assign natural numbers to number-predicates according to their complexity – 0 to the least complex, 1 to the second least complex, and so on *ad infinitum* – in such a way that given any number-predicate, there is a mechanical way of working out the corresponding

		0	1	2	.	.	.
0	—	no	yes	no	.	.	.
1	—	yes	yes	yes	.	.	.
2	—	no	no	no	.	.	.
⋮		⋮					

Figure 12.1

natural number, and *vice versa*. (There is no unique way of doing this. That is to say, there is no unique standard of complexity. One obvious criterion would be the number of symbols that the number-predicate involved. But it would still be necessary to discriminate between number-predicates with an equal number of symbols. At various points, arbitrary decisions must be taken.) Given that each number-predicate defines a set of natural numbers, we can then set up an 'infinite square' of *yeses* and *noes*, of the kind that occurred in Figure 8.4 (see above, Chapter 8, §3; this was used in the Cantorian proof that ℕ is smaller than its own power-set). On the top row of the 'square', there is an infinite sequence of *yeses* and *noes* registering whether successive natural numbers do or do not belong to the set defined by the least complex number-predicate, that which has been assigned 0. Similarly on successive rows. Figure 12.1 gives a different (arbitrarily chosen) example of what the 'square' might look like.

Suppose now that there *were* a decision-procedure for arithmetical truth. Then, in particular, there would be a decision-procedure for determining whether or not an arbitrary number-predicate was true of an arbitrary number. For example, we could use it to determine whether 'is odd' was true of 2 (no); whether 'is prime' was true of 13 (yes); whether 'is the sum of four squares' was true of 243 (yes); whether '< 2' was true of 821 (no); and so on and so forth. This in turn means that there would be a decision-procedure for determining whether there was a *yes* or a *no* at any arbitrary point in the 'square'. Now we already know how to use the technique of diagonalization to specify a set which is not on any row in the 'square'. We move down the 'square's diagonal' and write down a *yes* each time we arrive at a *no* and a *no* each time we arrive at a *yes* (see above, Chapter 8, §3). Call the resultant set *D*. Then there would be a decision-procedure for determining whether a given natural number belonged to *D* – yes, if it did not belong to the set listed on its own row in the 'square' (in other words, if there were a *no* at the relevant point on the 'diagonal'), and no, if it did (in other words, if there were a *yes* at the relevant point on the 'diagonal'). (For example, if the 'square' was as pictured in Figure 12.1, then 0 and 2 would belong to *D*, but 1 would not.) We are now on the brink of a contradiction, however. For it can be shown that, if there is a decision-procedure for determining whether a given natural number belongs to a given set of natural numbers in this way, then the set is 'tidy' enough for there to be a number-predicate that defines it. (The language of arithmetic is sufficiently rich for this to be so.) So what we have just shown is that there would be a number-predicate that defined *D*. But this contradicts the fact that *D* would not be listed on any row in the 'square'. So we must reject our original supposition. There could not, after all, be a decision-procedure for arithmetical truth. *Q.E.D.*

This way of proving Gödel's theorem is somewhat different from that originally adopted by Gödel himself. What Gödel did was to show how, given any sound axiomatic base *A* for arithmetic (sound in the sense that all its axioms are true), it is possible to exhibit a particular arithmetical statement *s*, such that *s* is true but cannot be proved using *A*. How is this done?

We can once again trade on the fact that it is possible to assign natural numbers to arithmetical expressions (including complete sentences) in various systematic ways. For then certain statements about natural numbers – arithmetical statements – can be seen to correspond to statements about their counterpart expressions – 'meta-arithmetical' statements. For example, the arithmetical statement that 90 < 300 might be seen to correspond to the meta-arithmetical statement that one expression is less complex than another. By following Gödel's ingenious strategy, itself based on the diagonalization exploited above, we can specify a complex arithmetical statement – call it *s* – that can be seen to correspond to the meta-arithmetical statement – call it *S* – that *s* itself cannot be proved using A. (It is as if s says, 'I cannot be proved using A.') We can then convince ourselves that s is true. For suppose it were false. Then S would have to be false too. In other words, it would have to be false that s could not be proved using A. That is, s – by hypothesis, false – *could* be proved using A. But this is obviously absurd: nothing false can be proved using a sound base. Having convinced ourselves that s is true, we can then conclude that S must be true too, which means that s cannot be proved using A. In other words, s is the sentence we were after.

The fact that s cannot be proved using A does not, however, leave us in any doubt as to whether to accept it or not. The whole point is that we also prove it to be true – using resources, obviously, that go beyond A. But what *exactly* are these resources? What extra key principles have we adopted that do not form part of the original base? It is in answering this question that we come to what Nagel and Newman describe as 'the coda of Gödel's amazing intellectual symphony'.[5]

Unless A is incomplete in a rather boring way, then we can show that a *lot* of our Gödelian proof that s is true can be recast and carried through using A. More precisely, there is an arithmetical statement – call it c – that can be seen to correspond to the meta-arithmetical statement – call it C – that A is consistent, and we can show that, using A, it is possible to get as far as a proof of

(1) s is true, provided that c is true.

This corresponds to the meta-arithmetical statement

(2) S is true, provided that C is true

or, in other words,

(3) If C is true (that is, if A is consistent), then S is true (that is, s cannot be proved using A).

We likewise can get as far as a proof of (3). What is it that enables us to go further, and to conclude that s *cannot* be proved using A, in other words that S is true and indeed that s itself is true? Answer: *the fact that we can recognize A's consistency*. And we can recognize A's consistency simply because we can recognize that all the axioms that comprise it are true. It is this which has no counterpart within A itself. There is a sense in which the heart of A's incompleteness lies in the fact that it cannot be used to prove c. (As it were: A cannot be used to prove its own consistency.) If c were added to A as an axiom, then the new axiomatic base $A + c$ could be used to prove first (1), then s itself (though $A + c$ would suffer from an analogous incompleteness of its own). So c is the key extra principle that we have adopted. But we need have no qualms about doing so. It simply registers our acceptance of A. The upshot of Gödel's theorem is therefore this: given any sound axiomatic base for arithmetic, our very recognition that that *is* what it is (a sound axiomatic base for arithmetic) propels us beyond it, and enables us to recognize the truth of arithmetical statements that it cannot itself be used to prove.[6]

Gödel's theorem raises a host of technical and philosophical questions. Much debate has arisen about what exactly its implications are for various different concerns. In the rest of this chapter, I wish to look into the question of what its implications are for our discussion of the infinite.

3 Hilbert's programme

I remarked in Chapter 9, §2 that Gödel's theorem seemed to dash any hopes of carrying out the two-part programme proposed by Hilbert six years earlier. That programme was: (i) to cast transfinite mathematics, including set theory and arithmetic, in an axiomatic form; and (ii) to prove its consistency in a finitary way. (I have expressed it somewhat differently from before.) It is true that Hilbert would not have accepted various parts of the second proof sketched above. For he would not have shared the background assumption that all set-theoretical and arithmetical statements can unproblematically be regarded as true or false. Ideal (non-finitary) propositions he would have regarded as neither. (They were formal devices brought in to facilitate proofs and to make for greater elegance and simplicity.) But there was enough here that he could, and did, accept, in a suitably modified form, to threaten both parts of his programme.

Why is (i) threatened? Because what Hilbert had envisaged – at least as a paradigm – was a single, complete axiomatization. And Gödel's theorem shows that nothing matches that paradigm. Any axiomatic base for transfinite mathematics must needs be supplemented. Not only that, but there will be one particular way of supplementing it that seems forced upon us; and this casts doubt on the idea that only finitary propositions genuinely describe mathematical reality. What will seem to force us to supplement the base in one way rather than another will be non-finitary reflection on the consistency of the base – reflection on the fact that, in the infinite landscape within which the base is located, there are no paths leading from it to each of two contradictory statements.

This brings us to why (ii) is threatened. How do we recognize the base as consistent? More to the point, how do we do this given that it contains ideal propositions? For Hilbert, there could be no question here of appealing to the *truth* of these propositions. We had to be able to prove that the base was consistent by treating it purely formally, as a system of meaningless symbols (as if it actually did consist of numbers, rather than statements). Our proof had to be non-ideal and non-question-begging. But Gödel's theorem shows in effect that no proof that the base is consistent can be recast and carried through using the base itself. And this seems to imply that no such proof can escape assumptions that are *at least as* extravagant as those which actually constitute the base, assumptions which are, indeed, ideal and question-begging.

True, none of this counts as a conclusive refutation of Hilbert's finitism, partly because of a degree of unclarity in how 'finitary' and 'ideal' are being understood. Gödel himself, in the very paper in which he presented his theorem, made this point, suggesting that there might be finitary methods of proof that could not be recast in the standard systems of arithmetic or set theory.[7] But this is surely cold comfort for the finitist. There is now a desperately delicate balance to be struck. A precise explanation of 'finitary' is required which both enables us to see why only finitary propositions directly hook up with mathematical reality and, at the same time, does not fall prey to Gödel's theorem.[8]

There is an issue here which extends beyond finitism. Suppose we revert to the view that every set-theoretical or arithmetical statement is either true or false, according to how things are in mathematical reality. We might still, of course, find ourselves confronted by a putative axiom such that we are not sure which of the two it is. For example, there are many statements that are candidates for being added to **ZF** as supplementary axioms to the effect that there exist large cardinals of different kinds. How can we tell which of these are true? Intuition offers little help (see above, Chapter 10, §4). One natural proposal is this: as many are true as it is possible to accept consistently. For we do have an intuition that the Set hierarchy extends as far as it possibly can; we are entitled to assert the existence of bigger and bigger Sets for as long as we do not lapse into inconsistency (say, by implying that there is a Set of all Sets). But how can we tell which of these statements can be consistently accepted? Not, it seems from Gödel's theorem, without relying on assumptions every bit as bold as those under scrutiny; and that will bring us right back to square one, trying to ascertain whether those assumptions are true.

There is also the question of how we tell whether **ZF** itself is consistent. It is all very well drawing 'V'-shaped diagrams intended to capture our intuitions about what Sets are like. But our most basic intuitions in this area have already proved unreliable. Confidence has in the past been misplaced. Is there any way of guaranteeing **ZF**'s consistency *without* relying on our intuitions about what Sets are like? It seems not, given Gödel's theorem. The lesson here is of a piece with the lesson that faced the finitist. If we are going to talk about the infinite in a mathematically precise way, then we really must see ourselves as *talking about the infinite*. And if we are going to ratify what we say, then we can make do with nothing less than insight into what the infinite is actually like. If all we have to go on are certain intuitions, then we must hope that those intuitions provide such insight – even if past experience should really make us wary of entertaining such a hope.

4 The human mind and computers

It has been argued by Lucas that one consequence of Gödel's theorem is that the human mind has mathematical powers beyond those of any possible computer.[9] We can gain a feel for Lucas' argument by reconsidering the dual presentation of Gödel's proof in §2. It was first argued, in outline, that there could not be a decision-procedure for arithmetical truth. This seems to point to inherent limitations in what any computer can do. For, assuming that computers operate by decision-procedures, it seems to follow that no computer can be programmed to discriminate between true arithmetical statements and false ones. At most it can be programmed to discriminate between the true and the false within a certain limited range of arithmetical statements. But here the second aspect of the theorem can be brought in. For given any limited range of this kind, we – human beings – can always recognize the truth of some arithmetical statement outside that range. This seems to clinch Lucas' contention. (Indeed there is no reason as yet to suppose that *we* cannot discriminate between true arithmetical statements and false ones quite generally, though, if we can, it may remain a mystery how.)

Lucas' argument has provoked much discussion.[10] A great deal of the discussion bears on issues that have begun to prove very pertinent to our own enquiry. Clearly one of the things that enables us to recognize the truth of a statement that is outside the range of a given decision-procedure or axiomatic base – in Gödelian fashion – is a kind of self-conscious reflection. Take **ZF**. Given that we accept all its axioms, we can, by self-consciously reflecting on this very fact, also come to accept its consistency, and it is this which enables us to see that certain set-theoretical statements are true even though they cannot be proved using **ZF**. We actually arrive at a deeper and more extensive understanding of what Sets are like. But such self-consciousness, I have argued, is of the very essence of set theory. It is the same self-consciousness which can, in ways that we have seen, unsettle us. In fact it unsettles us precisely when it is being exercised in *this* kind of way. We reflect on the consistency of **ZF** when we think that it is faithful to the intended interpretation of its language, in other words when we think that it faithfully describes . . . what? the full hierarchy of Sets? Here we are back in familiar deep water (see above, Chapter 11, §5). But if we can learn to exercise such self-consciousness without lapsing into incoherence, seeing **ZF** simply as a true account of what Sets are like, then Gödel's theorem offers a fascinating new perspective on what this may involve. It would be appealing if, as Lucas supposes, part and parcel of this was a clear demonstration of how our mathematical abilities outstrip those of any possible computer – precisely because they involve this kind of self-consciousness.

However, caution is called for. There are twin pressures that can be brought to bear on Lucas' argument, as it were from above and from below. This is not the place to go into their details, but I can sketch them.

The pressure from below turns on the thought that Lucas may have downgraded computers. Is it true that the only way that a computer could discriminate between the true and the false within a given range of arithmetical statements would be by operating a decision-procedure for truth within that range? Might there not be ways in which computers could be programmed to simulate human self-consciousness, perhaps with self-regulating devices, so that they could recognize the consistency of any principles they accepted?

The second pressure on Lucas' argument, the pressure from above, turns on the thought that he may have upgraded human beings. Certainly there is a sense in which, given any decision-procedure for truth within a range of arithmetical statements, we can specify an arithmetical statement outside that range which is true and which we can recognize to be true. But can we be sure that, in the relevant sense, we can be 'given' all such procedures? Might there not be a computer which somehow operated by a procedure that was too complex for any human being to grasp? And might such a computer not be just as good at discriminating between true arithmetical statements and false ones as we are? Perhaps our own powers of discrimination are limited. Perhaps they (in fact) correspond to a complex decision-procedure by which we nevertheless do not, and indeed could not, operate. Gödel himself had a related thought:

> There may exist ... a theorem-proving machine which in fact *is* equivalent to mathematical intuition, but cannot be *proved* to be so. ...[11]

5 Self-consciousness

Let us pursue the theme of self-consciousness. Gödel's theorem obviously helps to highlight how self-consciousness can be important in mathematics. It is through self-consciousness that we come to accept the consistency of principles we already accept. Indeed it is through self-consciousness that we encounter the question of their consistency in the first place (and go on to relate statements about their consistency to statements about numbers). For the question arises only when we stop thinking straightforwardly about numbers, or Sets, or whatever, and start thinking instead about (our own) *thinking* about numbers, or Sets, or whatever. We have to focus on the structure of our thoughts and on what it takes to express them in mathematical language.[12]

Recognizing ourselves as consistent involves seeing unity in the diversity of our thoughts, one-ness in their many-ness. This is why the self-consciousness that we are talking about here beckons paradoxes, at least in the set-theoretical case. We are always on the brink of recognizing a single unified subject matter for our thoughts: the Set of all Sets. What we may do, if we have a particular incomplete axiomatic base for set theory, is take all those Sets whose existence can be demonstrated using the base and then acknowledge the existence of a Set containing all of *these*. This may provide us with a natural way of extending the base, much like the Gödelian way of doing so, namely adding an axiom to the effect that such a Set exists; this is much like the Gödelian way of extending the base because it is tantamount to saying that the base is consistent (there being a Set which the base faithfully describes). But then the new, extended base will be liable to the same treatment. The full hierarchy must, in its infinitude, for ever elude our grasp. It must even elude our grasp when we reach that point of full self-consciousness at which we want to say that *it* is precisely what we have in focus.

6 Meaning and understanding

The discussion in the last section indicates something else that is integral to Gödel's theorem: a certain way in which infinitude resists capture in finite terms. The infinite richness of arithmetical truth is beyond the reach of any finite collection of arithmetical axioms. Likewise with the infinite richness of truths about Sets.

This brings us back to the original challenge presented by Gödel's theorem. How can we assimilate the truths of arithmetic or set theory, in order to learn the meaning of basic arithmetical or set-theoretical terms, if – what we seem forced to conclude – there is no suitable way of capturing these truths? However much we capture, it seems, there is always more to come. What then guarantees our concurrence about the more that is to come? How does it come about that we have a shared understanding of what natural numbers, or Sets, are like? Gödel's theorem presents a challenge very similar to that which the Löwenheim-Skolem theorem presented. It points to limitations in the Euclidean paradigm, and leaves us with a basic problem about meaning and understanding to resolve.

One way to meet the challenge would be to show that there was some way of capturing the truths of arithmetic and set theory *other* than by providing them with complete axiomatic bases. (Whatever the appeal of axiomatic bases, we must not regard them as sacrosanct. After all, people were engaged in arithmetic for thousands of years before any attempt was made to provide it with one.) Could we not take an incomplete axiomatic base, and add, as a general precept, that *whatever* principles we accept, or come to accept, are jointly consistent? This would be a single overarching expression of our self-consciousness. True, it would enjoy a kind of infinitude that prevented it from being cashed out as a further (single) axiom about what either numbers or Sets were like: it would be more like a general recipe for producing such axioms. But perhaps it would provide us with a way of capturing all arithmetical, or set-theoretical, truths. (Dummett, in his own discussion of this problem, comes very close to suggesting this, at least in the case of arithmetic.[13])

Certainly, if this suggestion were correct, the original threat posed by Gödel's theorem would be averted. But it is a real question how serious that threat is even if the suggestion is not correct. Suppose that there simply *were* no way of capturing all the truths of arithmetic or set theory. Would we then have a deep problem on our hands?

I think not. It is as well to consider what exactly the source of our worry is. We start with the idea that the meaning of an expression is a matter of how it is used. This in turn prompts the following thought: if the meaning is to be grasped, then something about how the expression is used, accessible to someone who does not already understand the expression, must serve to determine its meaning; and, whatever this is, it must, at least in principle, be describable in finite and non-question-begging terms. Otherwise, the thought is, we can have no real grip on how meaning can be reduced to use. To produce this finite, non-question-begging description would be, as I shall say, to *pin down* the expression's meaning. In the case of a mathematical expression, it would at least involve capturing all the truths of the formal theory in which the expression occurred. Hence our worry.

But these thoughts seem to me to be misguided. It is true, I believe, that the meaning of an expression is a matter of how it is used. But this does not mean that meaning can be *reduced* to use. Rather, meaning permeates use. We manifest our understanding of expressions in our use of them. We communicate with others by shared access to one another's use of them. There is nothing here to suggest that we should always be able to 'pin down' what an expression means.

In fact I do not believe that we can ever do this. Gödel's theorem helps to make graphic something about meaning which is there to be acknowledged anyway, something that I think deserves to be called the *infinitude* of meaning. The point is this. The meaning of an expression has infinite possibilities woven into it. Any expression can be applied in indefinitely many ways, for indefinitely many purposes, and to indefinitely many effects, whether literally, metaphorically, poetically, analogically, ironically, hyperbolically, precisely, roughly, or whatever. Nothing that we can describe in finite and non-question-begging terms is ever able to determine that full infinite potential. There is no legislating in advance for the possibilities of (creative) language-use that the infinitude of meaning affords. For example, there is no legislating in advance for the success of metaphorical uses, which may be contrived to describe situations completely unlike anything that anybody has ever encountered before. Such is the open-texturedness and versatility of meaning.

How then do people manage to grasp such meaning?

Well, how *do* they? They observe expressions being used. They try to see the point of the use. They try to use the expressions in the same way, under the guidance of promptings, corrections, and encouragement from others.

Yes, but if there is nothing in how an expression is used which they can have access to *before* understanding it and which actually serves to determine the expression's (full infinite) meaning, how does any of this help? Will they not be confronted by something which strikes them as being, at best, radically inconclusive and, at worst, so much incomprehensible babble?

Initially perhaps. But they eventually come to understand. It is true that this *can* seem quite mysterious. What we have to do, however, is to see it as perfectly natural. People just do have shared interests, and a shared sense of what is significant and of how things relate to one another. (These are partly innate and partly inculcated.) As a result, people are able to understand one another. They are able to see what other people are up to. They are able to grasp what expressions mean. In the mathematical case, there is no reason why being subjected to (some of) the truths of a formal theory – seeing how these truths are proved and the kinds of justifications that are proffered for them – should not give someone a sense of how to carry on, even though not all the truths have been, or could be, captured.

I am leaning heavily at the moment on some of Wittgenstein's later work on meaning and language use. At one point in his writing, he imagines an interlocutor suggesting that when he grasps what an expression means, then in a queer way its future use is in some sense already present. Wittgenstein replies that of course it is in *some* sense; the only thing wrong with what the interlocutor has said is the expression 'in a queer way'.[14] That captures beautifully the way in which we must stop finding mystery in what is in fact quite mundane. When somebody grasps the meaning of an expression, they simply come to understand it. They do not, mysteriously, gain insight into the future, even though the meaning of the expression is displayed in its continued usage, including its often unpredictable future usage.[15]

They do, however, acquire a *kind* of infinite power – the capacity to apply the expression in indefinitely many cases – a point of the sort made long ago by Aquinas (see above, Chapter 3, §2). Part of the infinitude of meaning has to do with the fact that it provides an unbounded framework within which we can interpret the world and describe its several finite aspects. Our understanding of things, including our understanding of our language, is not something finite we come across. It is our way of coming across finite things.

Of course, this is all somewhat removed from Gödel's theorem – though that last point calls to mind the idea of a general precept of consistency, guiding our encounters with, but resisting translation into, individual arithmetical or set-theoretical statements. What this discussion does do is take some of the sting out of Gödel's theorem, which now seems to give a technical twist to something non-technical and quite pervasive. We never learn meaning by seeing its full infinite potential being played out. What we see is always a finite portion of that. But the meaning is there (and not in any 'queer way'). What we must try to do is to discern it, to see what is going on. As we do so, the meaning becomes part of the framework within which we view things.

How, if at all, do we view the framework itself? Here again Gödel's theorem can help to cast interesting light. We view the framework by a kind of (introspective) self-conscious reflection.

That is a good cue for the topic of the next chapter.

CHAPTER 13

Saying and Showing

The one which is is both one and many, whole and parts, limited and yet unlimited in number. (Plato)

A question with no answer is a barrier that cannot be breached . . . [It] is questions with no answers that set the limits of human possibilities, describe the boundaries of human existence. (Milan Kundera)

All finite things reveal infinitude. (Theodore Roethke)

At various points in the last three chapters there was a sense of the ineffable in the background. Twice in particular it came to the fore. First, in Chapter 11, §3, I suggested that the sceptical or relativist reaction to the Löwenheim-Skolem theorem might be seen as (arising from) an ill-fated attempt to express an insight that somehow cannot be expressed. And later, in §5, I suggested that the same might be true of our continual allusions to the hierarchy of Sets – or indeed to the Set of Sets, if we treat the concept of the Set of Sets as a Kantian Idea of reason, and try to squeeze a regulative use out of it. Kant himself, when he allowed Ideas a regulative use, certainly seemed to be skirting the ineffable. It was very much as if there was something that he *wanted* to say (that the physical world exists as an infinite, unconditioned whole) though he had debarred himself from saying it, and this was his way of easing the tension.[1] We too *want* to say that there is a Set of all Sets. Is it satisfactory – is it, for that matter, legitimate – to speak *as if* there were such a Set, and then somehow to reconstrue what we have been saying so that it is not to be taken at face value? Maybe the right thing to do is to come clean, and to admit that we have an insight to which (we find) we cannot properly give voice.

My aim in this chapter is to lend some respectability to this suggestion. I shall try to make sense of the idea that there are certain insights that cannot be articulated; there are certain things that can be known though they cannot be put into words. Although this idea has so far been inchoate, there has already been enough to suggest an important link with self-consciousness (see above, Chapter 11, §5 and Chapter 12, §5). We have seen how self-consciousness in set theory can create an urge to say what we know we must not say. One of my goals is to substantiate this link between self-consciousness and ineffability. In the last two sections of the chapter I shall begin to address the question of how this all bears on the infinite.

I said in Chapter 9, when referring to Wittgenstein's work, that it can be divided into two phases. There I concentrated on the later phase. The main source for my thinking in this chapter will be ideas that dominated the earlier phase. My first task is to say something about the only philosophical work of Wittgenstein's, apart from one short article, to be published during his lifetime, namely the *Tractatus Logico-Philosophicus* (or the *Tractatus*, for short),[2] a remarkably austere and beautiful little book in which the whole of his early system of thought was encapsulated and set down in a series of elaborately interconnected aphorisms.

1 The saying/showing distinction in the *Tractatus*

Wittgenstein's aim in the *Tractatus* was to draw the limits of what can be thought and expressed. By carefully circumscribing the form of any possible language, he attempted to say what the world itself must be like in order for language to be able to 'picture' it at all. And by 'the world' he meant 'all that is the case',[3] all that can be thought or spoken of. He was led to a powerful vision, famous for its crystalline purity. The world was the totality of facts. Facts were determined by what he called states of affairs. States of affairs, each radically independent of all the others, were configurations of simple objects. Simple objects were what constituted the unanalyzable, ungenerable, indestructible substance of the world (*to apeiron*?).

It does not matter for our purposes how exactly all of this was to be interpreted. Suffice it to say that, in some sense, Wittgenstein was analyzing reality out – decomposing it into its ultimate logical constituents. And he analyzed language out in a parallel way. He held that simple objects were represented by what he called simple signs. That is, there was a matching of the most basic elements of reality with the most basic elements of language. And this meant that we could combine the signs to say, truly or falsely, how the objects were combined, or in other words what the facts were.

Wittgenstein was regarded by many as one of the chief architects, along with Russell, of what became known as logical atomism. Logical atomism, at least in Russell's case, was something of a reaction against the kind of idealism that had been prominent in Britain around the turn of the century in the works of such philosophers as Bradley and Taylor (see above, Chapter 7, §2). Such idealism had been heavily influenced by Hegel. In particular it had owed a great deal to Hegel's conception of the truly infinite (the metaphysically infinite). The world had been viewed as a complete self-contained indissoluble unity, no aspect of which made sense in isolation from any other. Much of logical atomism could be seen as a rather violent backlash to this. And certainly it seemed appropriate to cast Wittgenstein's vision in the same reactionary light. What can be thought and expressed, for Wittgenstein, was always a matter of isolated fact within the world, concerning independent simple objects. True propositions were accurate 'pictures' of such facts. It all seemed a far cry from Hegel's absolute whole, of which the truth itself was an essential and inseparable aspect. And yet . . .

Wittgenstein himself, in spelling out his vision, was treating the world as a limited and self-contained whole, whose parts were held together in unity. He wrote:

> Things are independent insofar as they can occur in all *possible* situations, but this form of independence is a form of connection with states of affairs, a form of dependence.[4]

He was talking about a kind of abstract, logical unity, holding things together. This did not determine what the facts *were*, but it did determine what they *could* be. It determined the limits of what could be thought and expressed.

What was going on here? Was it that his own brand of atomism somehow sanctioned his talking in these terms?

No. Notoriously, much of the *Tractatus*, when judged by its own lights, did not make sense. (This includes the sentence just quoted.) In drawing the limits of meaningful discourse Wittgenstein had stepped beyond them. In treating the world as a limited whole he had stationed himself outside it. The world, as a whole, was not itself anything *in* the world. But only what was in the world could be the subject of meaningful discourse. At the very end of the *Tractatus* Wittgenstein confessed that whoever understood him would recognize what he had been saying as nonsense: his reader must, so to speak, throw away the ladder after having climbed it. 'What we cannot speak about,' he wrote in conclusion, 'we must pass over in silence.'[5]

However – and this is the twist – what we cannot speak about, or what cannot be said, *can*, Wittgenstein maintained, be *shown*: the nonsense in the *Tractatus* had arisen from an attempt to put genuine insights into words. This distinction between what can be said and what can be shown – the saying/showing distinction – was a linchpin of the whole book. No feature of the world as a whole could properly be conveyed in words. The framework in which all the facts were held together was not itself a fact. Features of the world as a whole, its overall shape and form, were a matter not of its being *how* it was but of its being *however* it was. They were a matter not of what could be said but of what was involved in saying anything at all. They were what could be shown.

Moreover, it was what could be shown that principally concerned Wittgenstein. This was what he took to be truly important. His interest in drawing the limits of the thinkable and the expressible had more to do with what lay beyond them than with what lay within them (and of course, this included what was shown in the very drawing of those limits). For he believed that anything of *value* lay outside the world: value was a feature of the world as a whole, not of any fact about how things were. This belief came out in a compressed strain of mysticism towards the end of the *Tractatus*, in which ethics, aesthetics, and religion were brought together into a remarkable harmony. Wittgenstein held that being good (or, as he put it, being 'happy') meant viewing the world as a limited whole and adopting a certain attitude towards it, whereby its limits, so to speak, expanded. It was when we did this that we were shown what was of value. We were shown the world's beauty. We were shown its meaning. We were shown God.[6] Wittgenstein was deliberately laconic about all of this. To gesture towards it in the way that he did was as much as he was prepared to allow himself in view of its inexpressibility. But he famously wrote in a letter to Ficker:

> My work consists of two parts: the one presented here plus all that I have *not* written. And it is precisely this second part that is the important one.[7]

How does all of this relate to self-consciousness? To answer this question we must turn to Wittgenstein's treatment of 'the subject'. When I – the subject – view the world as a limited whole, I view it from my own particular point of view. Both its limits and its unity are shown in everything's being as it is *from that point of view, for me*. (What is of value is of value *for me*.) I am shown that the world is *my world*, and that I – the subject – am not myself anything *in* the world but rather its limit. This is connected with the way in which self-consciousness is not just consciousness of one particular element of reality. Focusing self-consciously and introspectively on myself, or on my point of view, I do not focus on one particular thing in the world that I come across or have access to; I focus on my way of coming across things; on the access itself. It is thus that I am shown the world *as mine*, and hence as limited and unified.

There is an analogy in the *Tractatus* that helps to make this clearer.[8] Suppose I consider my field of vision at any given moment. Then no matter how thoroughly I may describe what is in it, so long as that is all I do (that is, so long as I make no mention of anything outside the field), I shall not be able to say anything about the eye with which I see it all. For I cannot see my own eye. That there *is* such an eye, however, or rather that there is a point of vision at the edge of the field, is something I am 'shown' by how things are within the field, some of them being, for example, to the left, others to the right. And I am 'shown' this precisely when I stop concentrating on any particular thing *within* the field and indulge in a kind of self-conscious reflection on the structure of the field as a whole.

This is only an analogy, however. I can in fact say, with perfect propriety, that I see things with an eye. The full-blown saying/showing distinction, whereby there are things that cannot be said at all, emerges only when I pass from consideration of how things are in my field of vision to consideration of how things are, full stop – to consideration, in other words, of the world as a whole.[9]

2 The very idea of a saying/showing distinction[10]

My aim in this section is to move away from exposition of the *Tractatus* and to look into what we might call the very idea of a saying/showing distinction. The *Tractatus* will continue to be a touch-stone. But I shall be attempting to say, in very general terms and independently of the peculiar metaphyics of that book, how we might be justified in drawing such a distinction – or rather, how we might be justified in drawing a related, surrogate distinction. For I want to focus not on saying and showing themselves but on counterpart states of enlightenment. (This already involves a departure from the *Tractatus*.)

The surrogate distinction that I have in mind is the distinction between being in a state of enlightenment which can be put into words (as it were, having the answer to some question) and being in a state of enlightenment which cannot be put into words. If someone is in a state of enlightenment of the first kind I shall say that they know something that can be said. If they are in a state of enlightenment of the second kind I shall say that they are shown something. Henceforth in this book I shall only ever use the word 'show' and its cognates in this semi-technical (more or less Wittgensteinian) way. And I shall mean by 'the saying/showing distinction' this surrogate distinction.

The question is, are we shown anything? I shall try to argue that we are.

My starting point is the Kantian starting point: as metaphysically finite beings, we find ourselves cast into a world that is not of our own making. To know anything about that world we must let it impinge on us. We must be receptive. (See above, Chapter 6, §2.)

This means that, whenever we do know anything, there are two facets to our knowledge. On the one hand, there is that in our knowledge which is determined by what we receive (what is 'out there'). On the other hand, there is that in our knowledge which is determined by our own receptive capacities, which enable us to receive it (what is 'within'). Insofar as we can focus attention on these *severally* we can be in states of enlightenment of two different kinds. To be in a state of enlightenment of the first kind we must exercise suitable sensitivity to what lies outside us. To be in a state of enlightenment of the second kind we must be capable of a kind of introspective reflection on our own ability to exercise such sensitivity. In the first case we actually receive. In the second case we are self-consciously receptive.

It seems to me that this distinction exactly fits the saying/showing distinction as defined above. For putting a state of enlightenment into words means offering a potential receiver (anyone capable of understanding the words) a verbal picture of what one has received. One's offering mirrors one's reception. Only states of enlightenment which involve reception can be treated in this way. States of enlightenment which consist merely in being self-consciously receptive are not susceptible to being put into words. They indicate the limits of what can be put into words. To be in a state of enlightenment of this kind is, indeed, to be shown something.

There is much here that is still in line with the *Tractatus*. For what someone is shown, when they are self-consciously receptive in this way, is something about the framework within which they view the world. This determines the world's overall shape and form, *for them*. It is only how things are within the framework that can be said. Their own receptivity, the focus of their self-consciousness, is not a matter of how things are within the framework. It is a matter of what they are shown. Indeed we can follow Wittgenstein still further and say that they are shown each of the following: that they are not themselves anything in the world (remember the analogy with the field of vision); that they are a limit of the world; that the world is theirs.

But there is obvious cause for concern in this. It looks as if we have just tried to put into words what it is that they are shown, in the same way that Wittgenstein himself tried to put into words what is shown. Is the whole thing not self-stultifying? (Wittgenstein was prepared to admit that his own work was nonsense. Why think we can rest content with doing the same?)

I believe that these worries can be appeased. In fact we, unlike Wittgenstein, have not tried to put into words anything that is shown. What we have tried to put into words is what it is for somebody to be shown it. This distinction is crucial. Even if there is something self-stultifying about the project of expressing an inexpressible state of enlightenment, there is nothing self-stultifying about the project of describing somebody when they are in that state.

Maybe not, it will be protested, but the particular *way* in which we have tried to describe them surely does involve us in trying, at the same time, to say what it is that they are shown. For we have described them as being shown *that such and such is the case*. In schematic terms we have said something of the following form:

(1) *A* is shown that *x*.

How can *this* be legitimated?

I think we must construe (1) as having a misleading surface grammar (as

191

Wittgenstein himself might have put it at a later stage). Consider the following:

(2) *A* is shown something, and, *if* an attempt were made to put it into words, the result would be: *x*.

We can certainly say something of this form without ourselves endorsing the attempt, and without committing ourselves either to the truth or indeed to the meaningfulness of '*x*'. I suggest that we understand (1) as being equivalent to (2).[11] This means that (1) is altogether different from

(3) *A* knows that *x*.

Anything of *this* form can be true only if '*x*' is (meaningful and) true.

Let us take one of my earlier suggestions as an example. Suppose we describe someone as being shown that the world is their world. We need not think it makes sense to say that the world is their world (as opposed to yours or mine, say). And, even to the extent that we do think it makes sense, we need not think that it is true. All we are saying is that they have an inexpressible insight, and calling the world theirs is what we (or they) would be driven to if trying, forlornly, to express it.

Not that our only way of describing someone when they are shown something is this somewhat circuitous way. More direct and more natural ways are often available. What would be an example?

Let us suppose that what I said about meaning and understanding at the end of the last chapter is correct. It follows that anybody's understanding of their own language is a kind of receptive capacity. It enables them to receive (or to interpret) features of the world beyond. So by focusing on that understanding, and thinking self-consciously about the meaning of words, they come to be shown something about the framework within which they view things. The fact that they cannot say what they are shown is directly related to something else that I urged in the last chapter: that there is no way of 'pinning down' the meaning of any word or phrase. But we do have a way of describing *them*, in their inexpressible state of enlightenment. We can say that they are reflecting on what, say, 'green' or 'apple' or 'Set' means.

So although Wittgenstein tried to put what is shown into words, and was thus led to regard his own work as nonsense, we, in acknowledging a saying/showing distinction, need not try to do the same. Anything we say, in wielding the distinction, should meet the same *desiderata* as should anything we say in any other context. It should be meaningful, and true.

3 Wittgenstein's early views on the infinite

How does all of this bear on the infinite? One obvious route into this question is to see what bearing it had for Wittgenstein. Let us turn first to the *Tractatus*; later I shall cull relevant material from transitional work that he produced when his later ideas were beginning to take shape.

The Tractatus. *The metaphysically infinite*

The *Tractatus* itself did not contain a great deal that directly pertained to the mathematically infinite. (In an introduction that Russell wrote to the book he complained about its account of numbers precisely because it dealt only with finite numbers.[12]) The metaphysically infinite, on the other hand, was at the very core of the book; and it was directly related to Wittgenstein's saying/showing distinction. Reconsider the centrality of the idea of the world as a limited whole. In the closing pages of the book Wittgenstein identified what cannot be put into words (what can be shown) with what is mystical, and what is mystical with feeling the world as a limited whole.[13] The idea of a limited whole is deeply bound up with the idea of the metaphysically infinite. This was something that emerged at an early stage in the history of the topic with the Eleatics (see above, Chapter 1, §3). Parmenides, who was, I argued, one of the first thinkers to embrace the metaphysically infinite, did so by describing reality as a limited whole. Indeed he likened it to a finite sphere. The point of this, as I tried to explain, is that a thing's limits need not be imposed on it from without by something else; they may be imposed on it from within by its very own nature – so that the metaphysically infinite is bound to be limited, just like the mathematically finite. Parmenides' likening of reality to a finite sphere was of a piece with Wittgenstein's likening of reality to a person's field of vision. In each case we were being invited to understand the limits of the whole by analogy with spatial limits, and the whole itself by analogy with what was mathematically finite.

Is this too glib though? Even if these are only analogies, and even if we do not mind acknowledging tension between the two conceptions of the infinite, there is a whiff of the self-contradictory about all of this. How can it be the case that what is infinite has to be understood on the model of what is finite? Must we conclude that the idea of the metaphysically infinite is, after all, incoherent, and that there is no way of describing the whole *as* a whole?

Yes, I think we must – assuming that the metaphysically infinite really cannot be understood in any other way. The idea of the limited whole *is* incoherent.

But is this not an overly drastic conclusion to draw at this stage in the game? Is there not something unacceptably cavalier about simply abandoning the concept of the metaphysically infinite, given, as we have seen, just how crucial it has been to so much of the history of thought in this area? (It is hard to imagine that *nothing at all* was going on.)

There is indeed something unacceptably cavalier about simply abandoning the concept of the metaphysically infinite. But we do not have to. That is – putting the same point the other way round – we do not have to defend the concept against the charge of incoherence in order to retain it. The point is this. Suppose it *is* incoherent. Still there is an important role that it can play – precisely because of how it can be related, in Tractarian fashion, to the saying/showing distinction. This allows us to have our cake and eat it. We can describe someone as being shown that the world, as a whole, is metaphyscally infinite, without ourselves being committed to the coherence of the idea. Indeed we can describe *ourselves* as being shown this, without being committed to the coherence of the idea. However, I anticipate. These thoughts will be taken up in the next section and in the ensuing chapters (see especially Chapter 15, §§1–3).

Let us return to Wittgenstein's views. ('Views' is something of a misnomer. I mean the claims he made when trying to spell out what could be shown. These are not claims that, in a more authentic mode, he would have wanted to endorse.) As we have seen, he likened the subject's world – 'my world' – to a field of vision. This was bound up with his views about death and immortality. He held that my death was not a part of my world, just as the limits of my field of vision were not a part of *it*. Indeed, in the sense in which my field of vision had no limits (there were no limits within it), neither did my world; neither, if you like, did my life. But this had nothing to do with my being immortal. He likewise believed that my life had a kind of eternity that had nothing to do with my being immortal. He wrote:

> If we take eternity to mean not infinite temporal duration but timelessness, then eternal life belongs to those who live in the present.[14]

This was connected with his views about value and the meaning of life. The meaning of my life was quite independent of its duration. My life would have no greater meaning if its duration were infinite than it would if its duration were finite. Whatever meaning it did have it had *as* a limited whole, and in that sense its meaning did not lie within it, nor, therefore, within space and time.[15] True eternal life was a matter of metaphysical infinitude (wholeness, autonomy, integrity, being at one with the world) rather than mathematical infinitude (going on for ever). And, the concept of the metaphysically infinite's being what it is – more particularly, its being dialectically unstable in the way that it is – this was in turn a matter of finitude, *my* finitude. My life, my world, was a limited whole *because* it was mine. I *was* its limit. My finitude encompassed it.

There are two important connections here with existentialist themes, in particular those that surfaced in Sartre (see above, Chapter 7, §4). First, there is the nature of this finitude. It was no more a matter of mortality than the lack of limits *within* my life was a matter of immortality. It was a metaphysical finitude. Even if I were immortal, I would still be metaphysically finite and able to view my world as a limited whole, much as Sartre had insisted. My finitude provided my world with a framework and set its limits. For Sartre this was rooted in the fact that exercise of my freedom involved drawing these limits in, closing off certain possibilities. This is the second important connection with Wittgenstein. He too held that if I exercised my freedom in a certain way, then the limits of my world contracted – though I could also exercise it in such a way that they expanded. It was a question of what attitude I took (see above, §1). This waxing and waning of the world could be seen as its gaining and losing of meaning. And since the limits of the world determined not what the facts were, but what they could be, it is tempting to give this a further Sartrean twist by seeing the world's gaining and losing of meaning in terms of the opening up and closing off of possibilities. Certainly this has a ring of truth about it. Of all the marks of human finitude, the opening up and closing off of possibilities are as charged with meaning, surely, as any.[16]

The mathematically infinite

Let us now turn to Wittgenstein's transitional work, where he related these ideas to the mathematically infinite.

Possibilities are again important. In the *Tractatus* possibilities were part of what could be shown. The framework of the world determined what the possibilities were; how it was actually filled in determined what the facts were. Again, the form of objects determined what the possibilities were; how they were actually combined determined what the facts were. Knowing an object meant knowing the endless possibilities in which it might occur. The object *contained* those possibilities.

But this was all highly abstract. It was later that Wittgenstein began to probe the same ideas in more concrete forms. He turned his attention to space and time. He was impressed by the way in which any given spatio-temporal object, though only finite, nevertheless, in its very spatio-temporality, seemed to point beyond itself to the infinite possibilities of movement and reorientation of which it was capable. He was already shaping up to the idea that the proper use of such terms as 'infinity' was to generalize about possibilities of precisely this kind – the idea which dominated his later thinking on these issues (see above, Chapter 9, §3). But there was still a lingering Tractarian sense that such possibilities were the object of a special kind of experience, something that was ineffable. And what was true of the possibilities was true also of the infinite. The facts were so to speak finite, the possibilities infinite. The finite was rooted in what could be said, the infinite in what could be shown. In other words, although Wittgenstein was already regarding uncritical talk of 'the infinite' as a kind of nonsense, he had not yet fully surrendered the idea that it was a kind of nonsense that answered to something. In *Philosophical Remarks* he wrote:

We all . . . know what it means to say there is an infinite possibility and a
finite reality, since we say space and time are infinite but we can always
only see or live through finite bits of them. But from where, then, do I
derive any knowledge of the infinite at all? In some sense or other, I
must have two kinds of experience: one which is of the finite, . . . and one
of the infinite. And that's how it is. Experience as experience of the facts
gives me the finite; the objects *contain* the infinite.[17]

Again:

We are . . . only familiar with time – as it were – from the bit of time
before our eyes. . . . We are . . . in the same position with time as with
space. The actual time we are acquainted with is limited (finite). Infinity
is an internal quality of the form of time. . . . Doesn't it come to this: the
facts are finite, the infinite possibility of facts lies in the objects. That is
why it is shown, not described.[18]

A special concern with nonsense ran throughout Wittgenstein's thinking.
It constituted one of the most profound continuities between the earlier
phase and the later phase. In these quotations and the surrounding
material he was handling nonsense about the infinite in a way that bore
hallmarks of both: he was treating it with a distinctive combination of
antagonism and respect.

This, of course, makes the status of the work as transitional highly
significant. But there is another respect in which it is significant. For, as
Wrigley has argued, there were clear signs here of other preoccupations
from the later period.[19] I am referring to the problems about meaning and
understanding that were aired at the end of the last chapter. Wittgenstein
became greatly exercised by the question of how infinite possibilities of
meaning could be contained in finite portions of language use – a question
very similar to the one that was exercising him here, treated, I believe, in a
very similar way. He did not explicitly retain his saying/showing distinction
in the later work. (For example, he came to regard talk of an infinite reality
as straightforwardly ill-begotten.) But, for reasons that I tried to give in §2,
the view of meaning and understanding that derives from the later work
does lend itself naturally to the idea that, when we reflect on the meaning
of a word (with its full infinite potential), we are shown something.

This can itself be directly related to what we saw of Wittgenstein's later
views on infinity. On the one hand, he urged us to take uses of 'infinity' and
related phrases such as 'and so on' at face value. We were not to think of
them as pointing beyond themselves to anything. 'The expression "and so
on",' he wrote, 'is nothing but the *expression "and so on"* . . . [It] does not
harbour a secret power by which the series is continued without being
continued.'[20] On the other hand, because of their meaning, such expressions did seem to point beyond themselves. He wrote elsewhere as follows:

I believe that I perceive something drawn very fine in a segment of a series, ... which only needs the addition of 'and so on', in order to reach to infinity.[21]

There was a tension here, and, as so very often in Wittgenstein's later work, it was a tension which makes most sense if we think of him as probing what could be shown but trying to say only what could be said. He was probing meaning. He was trying, in as calm and sober a way as possible, to describe the linguistic practices in which the relevant expressions were anchored.[22]

4 The infinite and the ineffable

In this section I shall begin to explore the idea of an infinite framework shown to those capable of seeing and describing the finite elements within it. This idea is obviously extracted from what we have just been looking at in Wittgenstein. It seems to me that it gives us precisely the right way of seeing the relationship between the saying/showing distinction and the infinite. (What I have to say in this section will be largely programmatic. I shall try to substantiate it in the ensuing chapters, where various other Wittgensteinian themes will recur.)

I claim, first, that the saying/showing distinction, because of the way in which it enables us to have our cake and eat it, also enables us, in the same way, to address the deepest perplexities concerning the infinite that we have encountered in this book. To take a particularly stark and instructive example: we can solve what is, in my view, the most profound of the paradoxes of the mathematically infinite – (and one of the most revealing; it gave us an early and telling clue as to the nature of the link between the infinite and the ineffable) – namely the paradox of the Set of all Sets. The paradox is that we want both to affirm and to deny that there is such a Set. The solution is to say that we are *shown* that there is. Recall what this means. It does not follow from this, nor need we think it true, that there is in fact such a Set. As a matter of fact, we know that there is not; and we can say so. The point is this. When we self-consciously reflect on our set-theoretical practices we gain an ineffable insight. It is an insight into the unity of our subject matter (the infinite framework within which it lies). It is what leads us to believe that the claims we make are consistent (and hence, by Gödel's theorem, what leads us to endorse further claims). And *if* we were to try to give voice to this insight, then we should find ourselves proclaiming, falsely, that our subject matter formed a limited whole, or, more crisply, but still falsely, that there was a Set of all Sets.

This idea can be broadened.

Several times in this book, when discussing the mathematically infinite, I have talked about the 'truly' infinite; or I have qualified my references to the mathematically infinite in some other, similar way. This relates back to a view of Cantor's, namely that whatever is subjected to mathematical investigation, even if it is infinite in a suitably technical sense, is thereby seen to enjoy a *kind* of finitude: the truly infinite is that which resists mathematical investigation (see above, Chapter 8, §5). On this view, unless set theory is unacceptable for some reason, ω, $\mathbb{R}$, $_1$, 2^0, and all the rest are really finite.[23] What are really infinite are what Cantor called inconsistent totalities – totalities which, on the understanding of Sets canvassed above, do not really exist, for example Ω and the Set of all Sets.

It seems to me that Cantor was absolutely right to talk in these terms. What, after all, are the hallmarks of the (mathematically) infinite? They are: endlessness; unlimitedness; immeasurability; being greater than any assignable quantity. These point at least as much in the direction of inconsistent totalities as they do in the direction of what are technically known as infinite sets; and surely there is a sense in which they more truly apply to the former than to the latter. Similarly, the concerns and preoccupations that dominated the history of thought about the (mathematically) infinite until the time of Cantor had at least as much to do with inconsistent totalities as with infinite sets. In fact the history serves as a useful corrective against those inclined to say that, through the combined efforts of such people as Bolzano, Dedekind, and Cantor, we have at last been brought to see what infinitude really is: it is a property enjoyed by any set whose members can be paired off with the members of one of its proper subsets – as though *this* were the last word on everything that exercised Aristotle, Kant, Hegel, and all the rest. It is undeniable that this technical sense of infinitude bears directly on (and helps to dispel) a lot of earlier mystery about infinity, particularly, of course, mystery focused on the paradoxes of the infinitely big. But, especially if we want to remain faithful to Cantor's own vision, we do well to contrast infinitude of this kind with true infinitude. And this brings us back to the broadening of the set-theoretical idea to which I was alluding. What applies to the Set of all Sets applies, more generally, to the truly infinite. That is, *we are shown that the truly infinite exists*, though in fact (as we are bound to say) it does not.[24]

It does not exist because what does exist, in other words what is, is finite. What is is what can be encountered, addressed, attended to, grasped, managed, known, defined, handled, received, come across, given. It is what can, in some way or another, be *limited*. It would be an abrogation of the very concept of infinity to apply it directly to anything of this kind. (Compare this with the comments on the metaphysically infinite in §3, and with the idea, urged by Wittgenstein in his later work, that it is actually nonsensical to describe something as infinite; it is a violation of 'grammar'.) But we are shown that the truly infinite exists; we are shown that it embraces all of this, and holds it all together. (That is, we are shown that the truly infinite is finite. We are shown the finitude of infinity. Compare: we are shown the world as a limited whole.)

Saying and Showing

This is where the idea of an infinite framework can be brought in. It is as if whatever exists exists *in* some infinite framework (ultimately, the infinite framework of reality): but this always excludes the framework itself, which is at most shown to exist. If we were to look back over the course of our enquiry, we should notice variations on this theme scattered throughout it. Consider the following four examples:

(i) *Early Greek mathematics*: Here the theme was particularly striking. The objects of study in early Greek mathematics were always finite (lines, natural numbers, . . .), but their very study always presupposed an infinite framework (Euclidean space, $\mathbb{N}$, . . .). (See above, Chapter 1, §5.)

Comment: This set the pattern for nearly all of subsequent mathematics, as I have remarked a number of times. We saw something similar, for example, in the development of the calculus.

(ii) *Kant*: For Kant, space and time were part of the infinite framework through which we viewed what was beyond us; the things *within* space and time were all finite. Although he himself believed that we could talk about space and time with just as much right as we could talk about the things within them (in particular, we could say that space and time were infinite), there was a connection with the saying/showing distinction. For our knowledge that space and time were infinite was substantial, *a priori* knowledge, which meant that it was fundamentally different in kind from our knowledge about the things within them. The latter was the knowledge we acquired when we 'received' what was beyond us. (See above, Chapter 6, §3.)

Comment: I do not in fact believe that Kant was entitled to say as much about the framework as he did. The appeal to substantial, *a priori* knowledge, rather than vindicating him, actually served to emphasize that he was transgressing the limits of what could be said. (For amplification of this, see below, Chapter 14, §2.)

(iii) *Cantor*: As we have just seen, the theme was there in Cantor with his talk of inconsistent totalities.

Comment: This relates back to (i), and to the comment on it. At the one point in the history of mathematics where it seemed as if the infinite was at last being subjected to sustained mathematical scrutiny, still the pattern of early Greek mathematics was being repeated. This gives us another angle on the idea that it is inconsistent totalities that are truly infinite, not the so-called infinite sets that make them up.

(iv) *The* Tractatus: We came across the theme most recently, of course, in the *Tractatus*, where the only facts were facts about how things were

199

within the framework of the limited whole: there were no facts about the framework and the whole themselves.

What, finally, are we to say about those allusions to ineffability from the last three chapters that prompted this whole discussion in the first place?

Our references to the hierarchy of Sets, and to the Set of Sets, can indeed be seen as ill-begotten attempts to express the inexpressible. Likewise the excursions into scepticism and relativism, which, as we saw, go hand in hand with these. By self-consciously reflecting on the subject matter of our set theory, we are shown that the Sets which comprise it form a limited and determinate totality (the many forms a one). We are also shown that, because of this, there may be (must be?) loftier standpoints from which there are more Sets than these; our understanding is necessarily focused and limited. But we must not *say* this. It is incoherent to say that there are more Sets than *these*. These are all the Sets there are.

'What we cannot speak about we must pass over in silence.'

CHAPTER 14

Infinity Assessed. The History Reassessed

We are floating in a medium of vast extent, always drifting uncertainly, blown to and fro; whenever we think we have a fixed point to which we can cling and make fast, it shifts and leaves us behind; if we follow it, it eludes our grasp, slips away, and flees eternally before us. Nothing stands still for us. This is our natural state and yet the state most contrary to our inclinations. We burn with desire to find a firm footing, an ultimate, lasting base on which to build a tower rising up to infinity, but our whole foundation cracks and the earth opens up into the depth of the abyss. (Blaise Pascal)

What I want to do in this chapter is to continue what I was doing in the last section of the last chapter, but on a larger scale: that is, to reassess the history in the light of the ideas that have begun to emerge and, at the same time, to develop the ideas.

I shall continue to assume that the truly mathematically infinite is something that resists mathematical scrutiny. It does not follow that $\mathbb{N}$, $\mathbb{R}$, and suchlike are really finite. This only follows given the further assumption that they are susceptible to mathematical scrutiny. As we saw in Part One, there has been a good deal of scepticism about whether they are. The mere consistency of set theory, for example, would not be enough to assuage this scepticism with respect to $\mathbb{N}$. There remains the question of what any set-theoretical symbolism has to do with the natural numbers. If the answer is, 'Not enough to bring them under the control of the set theorist as a determinate, mathematically investigable totality,' then the natural numbers can continue to be regarded as providing a kind of paradigm of infinitude, in its truest sense – just as they have throughout the history of the topic. My assumption is not meant to prejudice any of these issues. It is just part of the model that I was beginning to develop in the last chapter.

One thing that the modern formalism does serve to emphasize is this: there is no single way of drawing together and subjugating everything in mathematical reality. For even if there is a determinate totality of natural numbers, and a determinate totality of real numbers, and a determinate totality of sets of natural numbers, and so on, still there is no determinate totality of *all* mathematical objects. (In particular, there is no determinate totality of Sets.) Related to this – (see above, Chapter 12, §5) – is Gödel's theorem. Whatever you are now focusing on, however much you have circumscribed, whatever you have got down on paper, or pinned down, you can be certain that there is more to come. That is the hallmark of the truly infinite. And this is one of the ways in which the modern formalism has made more graphic, and exacerbated, traditional perplexities surrounding the infinite. But, as I have tried to argue, it also gives us a handle on those perplexities in drawing our attention to the saying/showing distinction.

A note before I proceed: the main focus in this chapter is the mathematically infinite. I shall return to the metaphysically infinite, and to the relations between the two, in Chapter 15.

1 The infinite and the ineffable:
early Greek thought, medieval and Renaissance thought, post-Kantian thought

As so often happens in philosophy we find that we have been replaying themes from ancient Greece. I have already said something about Parmenides' relation to these ideas (see above, Chapter 13, §3). We can go back still further, to Anaximander and the Pythagoreans. There had been profound differences in how Anaximander and the Pythagoreans had regarded *to apeiron*. But both had seen it as a fundamental underlying principle of nature. It was required to make sense of, and it 'showed up' in, the finite things that we come across, even though it was not itself amenable to direct investigation. (The Pythagoreans assimilated it to what is female (see above, Chapter 1, §2). The reasons for this were many and varied. But it is worth considering whether the assimilation would have come so easily had there not been a sense of *to apeiron* as 'the other', and had the assimilation not been made by men.)

As the mathematically infinite gradually disentangled itself from this, in Zeno for example, it became an object of deep suspicion. It looked to be fraught with paradoxes and contradictions. But could it be escaped? In various ways the world seemed to exhibit infinite diversity, and, just as importantly, the infinite diversity seemed to be held together in unity. It was partly in response to this that Plato talked about Ideas. But precisely in talking about them he represented them as merely another part of the overall diversity. The unity remained tantalizingly ineffable – as he was in effect aware.[1]

Associations between the infinite and the ineffable were later made explicit by Neoplatonists, and by others whom they influenced. We saw both Plotinus and Nicholas of Cusa advocating a divine and transcendent infinitude that could not be captured in words, though it could be the object of a kind of direct, intuitive, mystical insight. But there was a paradox in their saying even as much as they did. This is one of the central paradoxes of thought about the infinite, as well as of religious thought. It is the kind of paradox that besets any attempt to come to terms with something that has been shown.[2]

The same ideas were played out in nineteenth- and twentieth-century thinking. The infinite again appeared as the object of a kind of inexpressible insight. Hamilton said that we could not think *what* it was, though we could think *that* it was. Bergson said that we had a direct, non-discursive access to it. There were many other examples that we saw. And it was widely accepted that as soon as we tried to talk about the infinite we became embroiled in a tissue of contradictions.

For Hegel (as indeed for Nicholas) this was because the infinite itself contained contradictions, which discursive thinking could not tolerate. By the infinite, Hegel (again like Nicholas) meant the metaphysically infinite. That, for Hegel, was what the infinite truly was. But the contradictions that arose when we tried to think about it discursively were played out in mathematically infinite terms – in a way that has interesting echoes in modern set theory. We would begin by thinking that we had a grasp of the infinite whole. Then what was within our grasp would appear, at the same time, to be finite, simply *qua* grasped. So it would yield to something more inclusive. Once again we would think that we had a grasp of the infinite whole. And so on *ad infinitum*. Each resting point was by its nature incomplete, unstable, temporary.

Later, the existentialists ventured expressions of the inexpressible. (Not that they would themselves have seen it in these terms.) In the last chapter I argued that the infinite is shown to us as finite. This is because of our own finitude. What we are shown is how that finitude structures and limits the world as a whole. Much of existentialist thought can be seen as an attempt to put this into words. I have already mentioned Sartre in this connection (see above, Chapter 13, §3). Heidegger went further. In trying to articulate the significance of human mortality he spoke of time itself, at the most fundamental level, as being mathematically finite. Compare Wittgenstein:

At death the world does not alter, but comes to an end.[3]

2 Aristotle and Kant: an unsuccessful compromise?

We can see a recurring tension in the history, a tension between what we want to say and what we are licensed to say. We want to acknowledge the existence of the truly infinite: but we cannot. Anything whose existence we can acknowledge we are bound to recognize as finite, on pain of contradiction and incoherence. The model sketched in the last chapter, with its appeal to the saying/showing distinction, is a response to this tension. But it may seem an extravagant and convoluted response. And it leaves us still unable to say what we want to say. It means conceding that there is no true infinity.

Can we not salvage from the history a more restrained, more cool-headed, and more satisfactory alternative? Reconsider Aristotle's actual/potential distinction, and, what was very closely related to this, Kant's advocacy (in deference to Plato) of a regulative use of our Idea of the infinite whole. Surely we can pluck from these the perfect compromise. We allow that the infinite exists, but we admit that it does not exist *now*, or at any other point in time; rather it exists over time. True, this is not faithful to the letter of what Kant said when he advocated a regulative use of our Idea. But it does seem to capture the spirit of what he said when justifying it: although it is incoherent to suppose that the physical world exists now, as an infinite whole, we know that, given any finite chunk of physical reality, there is always more to come.[4]

Is this a successful compromise?

No. I tried to indicate my disquiet about this kind of proposal at the end of Chapter 11. If we are so much as to recognize it as a compromise, rather than a way of just denying outright that the infinite exists, then it must fail. For unless we *now* have access to an infinite future (albeit perhaps only *as* the future, given our own immersion in time), we have no way of understanding how anything that exists over time can exist as an infinite whole.

This is not meant as deprecation of either Aristotle or Kant. There is still the question of whether they intended their proposals as such a compromise. I shall return to this question in §4. I do think, though, that Kant, like the existentialists, was involved in a continual, deep-seated attempt to put inexpressible insights into words. He argued that, as metaphysically finite beings, we supplied ourselves with a framework through which to view things – a framework which, in some sense of 'could', could just as well have been otherwise and which was itself metaphysically finite.[5] But even by his own lights, there was no legitimate way of getting outside the framework to talk about it in this way. Nor, if it was spatio-temporal in the way that he believed, did he have any right to talk about space and time themselves as wholes, worse still as infinite wholes. Nor was there any legitimate way of talking about the framework's filling, as a whole.

On this last point, however, he escaped reproach. Kant himself was especially and effectively militant against the urge to talk in that way. For it was none other than the urge to apply the Idea of the infinite whole non-regulatively to physical reality.

3 The empiricists: an uncompromising success?

Curiously, it is the empiricists who begin to look like the heroes in this drama. At least they were not ushering the infinite in under false pretences. Aware of the special difficulties that the concept of infinity posed on their own empiricist principles, they refused, mostly without compromise, to have any truck with it at all. And this was commendably ingenuous.

Nevertheless, to reject the concept as completely useless, which is what empiricism in its most extreme form sanctions, is to go too far. The concept does have a role to play – not least, as we have seen, in enabling us to describe people when they are in certain inexpressible states of enlightenment (though not only in that way; see below, §4 and Chapter 15, §3).

I said at the end of Chapter 5 that the infinite (more strictly, the concept of the infinite) helped to signal the failure of empiricism. We are now better placed to see why. The empiricists, with their special commitment to experience, regarded objects of knowledge as isolable components, packaged by the different senses. (Note the close affinities with logical atomism (see above, Chapter 13, §1).) But this left them with no satisfactory account of how the components were held together in unity. Hume famously confessed, in an appendix to his great work *A Treatise of Human Nature*,[6] that he could not see any way round this problem. Now it was part of the genius of Kant that he paved a way – a way that enabled him to salvage many of the empiricists' insights, in particular the insight that such unity was not itself a simple object of experience. To echo Strawson's memorable phrase: there was a unity of experiences without any experience of unity.[7] But Kant himself went too far, as I have just said, in trying to describe this in terms of our actually imposing a unified framework onto our experiences. There is no way of saying where the unity comes from, nor of describing it directly. There is no way of describing the one-ness of the many. What we can say is that we are *shown* such unity, through self-conscious reflection. This is where the concept of infinity has a role to play. The unity of consciousness that we are shown is the obverse of the unity of the world – 'our world' – conceived as an infinite whole. (For amplification of these ideas, see below, Chapter 15, §3.)

In their silence over the infinite (or at least, the infinitely big) the more robust of the empiricists displayed a tenacity that was entirely proper, in its own way. But there was more to be said, and it was a major problem for them that they had deprived themselves of the means of saying it. Still, if any group in the history can be singled out as attempting to express only the expressible, then this was that group.

205

4 The Wittgensteinian critique. Aristotle and Kant vindicated?[8]

The (truly) infinite, I have maintained, does not exist. That rules out at least one simple use of such terms as 'infinity': to describe something as actually being infinite. But viable uses remain. I have tried to highlight one such use: to say what it is for people to be shown certain things (though not, of course, to say what they are shown). But there is also the much more straightforward and familiar use canvassed by Wittgenstein in his later work: to characterize the form of finite things and to talk about the endless possibilities that finite things afford (see above, Chapter 9, §3).

We can say, for example, 'There are infinitely many natural numbers,' or, 'There are infinitely many ordinals,' meaning: however many natural numbers we have counted we can count more; or, whatever totality of ordinals we have characterized we can characterize a more inclusive totality. For these are ways of saying how, in the nature of the case, certain possibilities always obtain. They are not claims about objects – sets – that are actually infinite (though we may be happy to license their being couched in that way as a *façon de parler*). Similarly we can say that space and time are infinite, provided that we understand this in the way sanctioned by Wittgenstein (see above, Chapter 9, §3).

For Wittgenstein, of course, it was not just false to say that the (truly) infinite exists, it was incoherent. That is, it was nonsense, a violation of 'grammar', a mishandling of the language. You could not be said to have understood what kind of 'tool' the term 'infinite' was unless you realized that it could never be directly applied to anything. We have every reason to follow Wittgenstein here. That the infinite does not exist is indeed a matter of 'grammar'. It cuts that deep.

The Wittgensteinian critique also casts further interesting light on the history. For one thing, the empiricists appear once again to have been vindicated. When, in their less obdurate moments, they allowed themselves to wield the troublesome vocabulary of infinity, it was in precisely the right kind of way – to convey possibilities of repetition, the unimaginability of limits, and suchlike. (Consider Locke's 'negative ideas'.) By the same token it is the rationalists who begin to look like the real villains of the piece. For it was there in the history that we saw the boldest and most unabashed commitment to the existence of an actual infinity, a morass, as it now appears, of conceptual confusion.

But matters are not so simple. The empiricists, in backing down as far as they did, were already compromising an authentic empiricist account of the mind's conceptual apparatus. It was supposed to consist of ideas, copied, image-like, from previous direct experience.[9] The rationalists, for their part, in insisting that our idea of the infinite could not be traced back to experience or to imagination, were, so I would argue, presaging the kind of demarcation between different states of enlightenment that underlies the saying/showing distinction.

More interesting is the light that Wittgenstein's critique casts on Aristotle and Kant. In §2 I rejected a proposal deriving from their work. It was a proposal that was designed to enable us to assert the existence of the truly infinite. But did either Aristotle or Kant intend anything like this? Perhaps they themselves were only adverting to the possibilities inherent in the form of things.

Kant certainly was. Exegesis of Aristotle is more delicate. I shall turn to Kant first.

Kant: Kant's idea was that all finite things in space and time were conditioned. For example, he believed that nothing happened without a cause. But this only meant that, in the nature of things, a condition could always be sought and found. A regress of conditions was thus '*set us as a task*', but it was not 'already really *given*'.[10] We could use our Idea of the unconditioned whole as a rule, to explain how to set about this task and to enjoin ourselves to do so. But this was not to say anything about how things really were. It was certainly not to ascribe an actual infinitude to anything.

Comment: We might wonder whether Kant's claim that space and time were given us as infinite wholes could also be vindicated in these terms – as meaning that we knew, *a priori*, that there would always be space and time enough for the conditions of anything to be found. It seems clear, however, that Kant intended something more robust. For he held space and time to be infinite in a sense in which the physical world was not. If we adopt the suggestion just made, there is no accounting for this difference.

Aristotle: Aristotle certainly said that the infinite exists over time. But he may have understood this in accord with the kind of proposal that Wittgenstein was making. That is, he may not have been granting the existence of 'an infinite reality'. He may have been saying that certain processes, by their very nature, would never give out. It is in any case a real question just how straightforwardly his temporal allusions were to be taken. Consider, for example, his claim that matter was infinite by division. It would be a rather crude reading of this claim, or at any rate it would seem to be, to suppose that he was actually anticipating a non-terminating process of cuts (even if they were naturally induced cuts). Was he not saying that, in virtue of its form, a body, however many times it had been divided, *could* always be divided further?[11] (This is a genuine question. The crude reading is not to be sneered at.[12] Temporality and possibility were intimately related for Aristotle. Much turns on the exact nature of the relation.)

Comment: Even if Aristotle did hold the position discredited in §2, and even if he did face an insurmountable problem about the future, rather like the problem that he already faced about the past (see above, Chapter 2,

§5), his contribution to the history of thought about the infinite is scarcely diminished. For the actual/potential distinction itself remains intact. (So does much that went with it. Aristotle could still maintain that nothing spatial was actually infinite, either by addition or by division, and that space, time, and matter were all potentially infinite, by division. What he could no longer do was deny the actual infinitude of time, or indeed of number; not of number, because he held that time itself was 'a kind of number'.[13]) The actual/potential distinction has been of unrivalled importance as a guide to understanding the infinite. This is not least because of how, *via* such intermediaries as the medievals' categorematic/syncategorematic distinction, it paves the way for what we find in Wittgenstein.

Let us finally consider an objection to the Wittgensteinian critique, suggested by this very fact, namely that it faces an analogous problem to the one posed in §2. For are we not effectively saying that the infinite exists over an infinite range, not of times but of possibilities – for example, when we say that a particular body is infinitely divisible – and must this not be understood in terms of access which we *actually* have to that range?

Wittgenstein was alive to this objection.[14] He argued that it rested on a deep confusion. The confusion was to think that possibilities were things that enjoyed a (shadowy) kind of reality. This was something he railed against many times.[15] Again I think we should follow him. Once the confusion is cleared up, the idea that we are granting the infinite an existence over a range of possibilities, or indeed an existence of any other kind, seems crazy. To say that a body is infinitely divisible is to say something about the form of the body, as it actually is, in its finitude – much as, when we say that it is possible for a particular peg to fit into a particular hole, we are saying something about the respective sizes and shapes of the peg and the hole. It is no more mysterious than that.

5 The impossibility of an infinite co-incidence, and the law of the excluded middle[16]

There are infinitely many ordinals. This does not mean that there is an infinite totality of ordinals. It means precisely that there is no totality of ordinals. Given any totality, there is another that is more inclusive.

It may be that there are infinitely many natural numbers, in that same true sense. This depends on whether $\mathbb{N}$ is a legitimate object of mathematical study or not. If it is not, in other words if there *are* infinitely many natural numbers, then they comprise all the ordinals. We cannot make sense of an ordinal that succeeds them all if they do not themselves form a determinate totality. In particular we cannot make sense of ω if they do not do that. (ω has after all been construed as the Set of natural numbers.) But be that as it may, there are, to repeat, infinitely many ordinals.

This means that they are a many that cannot count as a one. No construction, no enumeration, no survey can exhaust them all. There are always more to come. Note the deep temporal metaphor that pervades this – the sense of a never-ending process successively yielding them (see above, Chapter 10, §1). Intuitionists, for whom this is more than just a metaphor, deny that the process can ever get beyond the natural numbers.[17] But again I want to focus on an issue that arises irrespective of that.

How are these temporal overtones to be understood, in the light of the Wittgensteinian critique? Not in such a way that the never-ending process can already be regarded as a whole. The idea is rather that the process, by its very nature, can never give out. But this must make us balk, as it made the intuitionists and Wittgenstein balk, at the idea of an infinite co-incidence among the ordinals. How can it just *happen* that the ordinals are all alike in a given respect? For them all to share a certain property, or for them all to lack a certain property, is for there to be some way of seeing in advance that they *must* do so. It is for there to be some principle whereby the process is constrained to carry on in a certain way. For example, there is the fact that it is constrained to carry on at all: we know that every ordinal must have an immediate successor. But since the process never terminates, it cannot just *turn out* that each ordinal is like every other in some particular respect.

One thing that follows from this, as Brouwer, Wittgenstein, and Dummett have all argued, is that we cannot uncritically assume the so-called law of the excluded middle. This is the law that, given any proposition, either it or its negation is true. For example, suppose that φ is some mathematical property. Then we cannot assume, without further ado, that either there is at least one ordinal which has φ (call this proposition 'Some-φ') or this is not so: none of the ordinals have φ (call this proposition 'No-φ'). This is because, if we do not yet have any way of seeing which is true, then to make the assumption is to register indifference about whether there *is* a way; and this, at least in the case of No-φ, is to acquiesce in the possibility of an infinite co-incidence. What it comes to is this. We are refusing to assume, without further ado, that either the process will eventually yield an ordinal which has φ or there is some general principle to rule this out.

Not that we are envisaging a third alternative. A third alternative *can* be ruled out. It is impossible for both Some-φ and No-φ to be false. This is because, if Some-φ is false, then No-φ, for that very reason, is true. For No-φ is the negation of Some-φ. We are not actually rejecting the law of the excluded middle then. That is, we are not actually specifying or envisaging a counterexample to it. What we are doing is holding back from an indiscriminate application of it, underpinned by the idea that there may be infinite co-incidences.

There are obvious affinities here with Kant. He would not sanction the assumption that the physical world, as a whole, was either infinite or finite in various specified respects. At the same time he was at pains to emphasize that he was not actually rejecting the law of the excluded middle.[18] However, Kant's position was more straightforward than ours. The assumption that he was refusing to make was not an assumption that involved a proposition and its negation. There was, in his case, a third alternative that he was at perfect liberty to embrace, as indeed he did, namely that the physical world did not exist as a whole. It would be appealing if we could see our own wariness in the same rather innocuous light – so that all we were really doing was refusing to commit ourselves to the existence of Ω, the Set of all ordinals. But neither Some-φ nor No-φ makes any reference to such a Set. No-φ really is the negation of Some-φ.

Of course, it is tempting to think that somehow it is not, that Some-φ and No-φ might both be false, and that this in turn is why we are being wary. It is tempting, in other words, to think that the following might be the case: although there is no general principle to rule out any ordinal's having φ (so No-φ is false), *in fact* none of them does (so Some-φ is false as well). But this is precisely to envisage an infinite co-incidence, the very possibility that we are supposed to be rejecting. Our position is thus a delicate one. We must demur when others uncritically countenance the assumption that either Some-φ or No-φ is true. But we have no way of responding to the question, 'Why? What can have gone wrong? In what circumstances would this assumption be false?' It is as if we are keeping faith with a kind of inexpressible scepticism – again.

In fact that is, I believe, what we are doing. Our position here is very much like that of the sceptic in Chapter 11, responding to the Löwenheim-Skolem theorem. The point is this. We are shown the existence of Ω, as if spread over infinite time. We are shown the (infinitely) many ordinals together forming a one, unified by our very ability to speculate and make generalizations about them. (This unity is not something that is within our ordinary mathematical access, in the way that each individual ordinal is. It is a feature of the access itself.) But what we *say* about the ordinals must be conditioned by the fact that we are inescapably immersed in time and can only make sense of what is accessible from a vantage point within time – however literally or metaphorically these temporal references are to be taken. (That is after all part of what it is not to be able to recognize a Set of all Sets.) This is why we must deny the possibility of an infinite co-incidence. A generalization about the ordinals can only be true if there is something that establishes its truth in advance (in time). We are nevertheless *shown* that an infinite co-incidence is possible. We are shown that our conception of a generalization is unduly limited by our time-bound perspective and that No-φ, as we are constrained to understand it, is not the negation of Some-φ. *This* is what makes us wary of assuming uncritically that either Some-φ or No-φ is true, though we have no way of articulating how the assumption could be false. There is a kind of interplay between the two sorts of enlightenment – between what we are shown and what we know and can say – an interplay that trades on their very tension.[19]

Not that we are compromising what we say in acknowledging this. It is not as if we are trying to say what we are shown. On the contrary we are being particularly cautious in our efforts to say only what can be said. We are refusing to assume what we have no reason to assume – though we equally have no reason to reject it, and indeed can have no reason to reject it. The link with what we are shown is simply that our caution derives from a state of enlightenment that cannot itself be put into words.

6 A problem for intuitionism[20]

Much of the previous section owed a great deal to the intuitionism of Brouwer. I shall close this chapter by returning to a problem for this position to which I drew attention in Chapter 9, §4.

For extreme intuitionists there is no question but that there are infinitely many natural numbers, in the truest sense – and no ordinals beyond them. Mathematics is a matter of what can be 'constructed' in time. We can never reach the point where we have constructed all the natural numbers.

I say 'for extreme intuitionists'. Interestingly, this does not include Brouwer himself. He acknowledged the existence of countable ordinals beyond the natural numbers – ω, $\omega + 1$, and so on. For him the truly infinite did not arise until $_1$. He thought that the operation whereby successive natural numbers were constructed, namely that of adding 1, itself gave rise to ω, to which it could in turn be applied.[21] I find this hard to reconcile with his own conception of mathematical construction, as grounded in our experience of time. If the construction of the natural numbers is literally never-ending, how are we to explain the addition of 1 to ω? But be that as it may, Brouwer did share the fundamental intuitionistic tenet that we can never reach the point of having constructed all the natural numbers.

We can, however, construct *each one*, however large. This is because the 'can' is a 'can' of principle, not of practice. More precisely, it is a 'can' of temporal immersion. What counts is what is or is not possible given (only) that we exist in time. Mental and physical limitations are by the by. This also applies to calculations. For example, we can, in this same sense, calculate the product of a pair of trillion-digit numbers, even though in practice this is out of the question. (Our lives are too short, our memory and powers of attention are too feeble, we do not have enough paper to record the steps in our calculation, we *need* to record the steps in our calculation, *et cetera*.) Intuitionists are concerned with a distinction

211

between the temporal and the atemporal. They are not, as such, concerned with a distinction between the little and the large.

But now the following problem arises. Given that practical constraints are not at issue, it seems that we can, after all, construct (survey, inspect) all of the natural numbers in a finite time. We just need to deal with each twice as quickly as its predecessor. We can then deal with all of them in only twice the time it takes to deal with 0.

Of course, this is not *in fact* possible. Sheer sluggishness will eventually thwart us – or, if not sluggishness, then cack-handedness, as we try to write smaller and smaller, or, if not these, then the minute physical constitution of the universe. But one thing that will not thwart us, it seems, is our immersion in time. (I ignore the point that time itself might be quantized. If it were – even if it necessarily were – intuitionists would surely be prepared to recast their philosophy of mathematics in terms of a hypothetical, continuous time. For what matters, I take it, is what our immersion in time rules out on essentially *a priori* grounds – though it is a nice question, what question-begging this might involve.)

This problem is not just a problem for intuitionists. It is a problem for anyone (such as Aristotle, Wittgenstein, and perhaps Dummett) who believes that (i) there are infinitely many natural numbers, and (ii) there is therefore a sense in which they can never all be constructed (surveyed, inspected) which is *different* from the sense in which all those with fewer than a million digits, say, can never be constructed (surveyed, inspected).

Of course, someone who denied (i) could avoid the problem. A particularly interesting example of this would be someone who thought that all countable sets ($\mathbb{N}$, ω^2, ω^ω, and all the rest) were really finite – precisely because it is possible to perform countably many tasks in a finite time. They could still maintain that there are infinitely many ordinals, and it would be provably impossible to pin an analogous objection onto them. There is no Zenonian procedure whereby it is possible to perform uncountably many tasks in a finite time.[22] (Note: although Brouwer himself was an example of someone who thought that all countable sets were, in this sense, finite, the problem was as much his as anyone's – precisely because he did not think that it was possible to construct all of the natural numbers in a finite time.)

Again, someone who denied (ii) could avoid the problem. For them, the difficulties concerning the infinite would be essentially no different from the difficulties concerning the very large. Of course, if they thought that an intuitionistic wariness was appropriate in response to difficulties of the first kind, then they ought to think the same thing about difficulties of the second kind. For example, given some suitably colossal number, they ought not to assume uncritically that either it is prime or it is not prime. I do not myself believe that this position is tenable. It was a crucial feature of the wariness that I was advocating in the previous section that it owed something specifically to the infinite. (Wittgenstein, incidentally, although he did take practical considerations very seriously in his philosophy of mathematics, is not an example of someone who denied (ii). This should be clearer in a little while.)

Anyone who believes both (i) and (ii), however, is faced with a problem, which we need to address. As we do so, we shall also get a firmer grip on the paradoxes of the infinitely small. For we are confronting the whole question of whether it is possible to perform infinitely many tasks in a finite time.

Consider first the following response to the problem:

It is, admittedly, only a practical impossibility for us to construct all the natural numbers. That is, it is an impossibility due to mental and physical limitations. These limitations nevertheless lack the *specificity* of those which prevent us from constructing all the natural numbers with fewer than a million digits. In the former case it is a matter of there coming a point – some point or other – beyond which we cannot exert ourselves. In the latter case it is a matter of our not being able to exert ourselves beyond a *particular* point. If we could work ten to the million times more quickly than we do, the latter task would no longer be beyond us. The former task, on the other hand, still would be. The former task would remain beyond us however far we extended our powers and abilities, so long as they were still subject to *some* limitation. This is enough to justify regarding the difference between the two cases as a difference of kind, and thus to vindicate (ii) – and, for that matter, (i).

This is a powerful response. We may wonder whether it provides rationale enough for refusing to license ℕ as a determinate, mathematically investigable totality, in other words whether it really does vindicate (i). But certainly it does justice to our sense that the extra difficulties involved in constructing all the natural numbers, as compared with constructing all those with fewer than a million digits, is not simply a matter of there being more of them. The problem, however, is that although this will satisfy some of those originally threatened, it cannot satisfy the intuitionists. For it involves a crucial concession: that our inability ever to construct all the natural numbers is not just a matter of our being immersed in time. The infinitude of the natural numbers is no longer being understood in terms of pure temporal structure. (Aristotle, for the same reason, would not have been satisfied. Nor would Wittgenstein, for whom the idea of performing infinitely many tasks in a finite time needed to be exposed as positively incoherent.)

Let us call any story in which infinitely many tasks are performed in a finite time a 'super-task story'.[23] Then one thing is surely beyond dispute: that logically consistent super-task stories are there for the telling. In one such story someone constructs all the natural numbers in a minute. They spend half a minute constructing 0, a quarter of a minute constructing 1, and so on *ad infinitum*. We can say how long it takes them to get to any given number (25, say); and we can say how long they spend constructing that number (in the case of 25, just under a microsecond).

But logical consistency does not guarantee coherence – a point that intuitionists are especially keen to emphasize.[24] (A cornerstone of their philosophy of mathematics is that consistency can fall short of genuine meaningfulness.) And it is hard to shrug off the feeling that super-task stories, for all their consistency, are not in fact coherent. Some of the paradoxes of the infinitely small reinforce this feeling. Take the paradox of the divided stick (see above, introduction, §1). If we allow that a stick can be divided into infinitely many pieces in this way, then we are left needing to make sense of the idea of an infinitesimal slice of the stick. An infinitesimal slice of the stick would seem to be a surface that continues to exist even when whatever it is a surface of has been destroyed. It may be said that this is an unfair characterization, and that we can make sense of infinitesimal slices in terms of what happens in certain regions of space. But it is far from obvious how. In any case, we also need to make sense of the idea that infinitesimal slices can make up the stick. That remains deeply mysterious.

But the possibility that super-task stories are *not* coherent will not – yet – help the intuitionist. For there can be no non-question-begging way of explaining the incoherence if all we have to appeal to is pure temporal structure and the fact that we are immersed in time. We need some independent leverage. But what?

It is not enough to throw out a challenge in the way that Wright does:

> How, if we were suddenly enabled to perform infinitely many tasks in a finite time, could we know that we were?[25]

For this invites the simple response, 'By doing it.' Nor is it enough to point out that we cannot *imagine* doing it. This is all too easily explained. Our imagination is subject to the same kinds of limitations, and is sluggish in the same way, as our executive abilities. In fact fully to imagine constructing all the natural numbers would be tantamount to actually doing it, given the understanding of construction intended by Brouwer (see above, Chapter 9, §1).

I see no alternative but to turn back to the Wittgensteinian critique and to look again at the 'grammar' of 'infinity'. Somehow we have to see such sentences as, 'This stick has been infinitely divided,' and 'I have just constructed all of the natural numbers,' as abuses of grammar, misappropriations of the language.[26] These can be subtle and well-disguised. They need not be syntactically unacceptable. Nor, as we have seen, need they be logically inconsistent. For example, the following sentence would also have to count as improper by these lights:

I spent half a minute constructing 0, and, each time I had finished constructing a number, I then spent half as long constructing its successor, until a minute had elapsed.

But what is wrong with this? It is certainly possible for me to spend half a minute constructing 0. It is also possible for me to spend three quarters of a minute constructing first 0, then 1. Can this not be continued *ad infinitum?*

It can, but only in the sense that there is an endless series of possibilities. There is not one possibility involving an endless series. There is a basic misunderstanding involved in trying to collapse all these scenarios, of indefinitely increasing complexity, into a single scenario.[27] It literally makes no sense to describe anything as infinite in this or that respect. We can only use 'infinity' to describe the endlessly nested possibilities that (finite) things afford. That is the only use that we have learned to correlate with any aspect of our experience. It is the only use that can serve to direct our attention to anything. (Even if we had empirical evidence that there were infinitely many stars, say, this would have to be understood in terms of the infinite possibilities that were written into what we could encounter in experience: the evidence would have to be that, if we were to travel further and further away from some point, we should always be able to find more stars.)

Think of it like this. If it did make sense to say that I had just constructed all of the natural numbers in a minute, by the Zenonian procedure, then it would also make sense to say this: while I was constructing them, my constantly increasing speed of performance meant that time seemed to be going more and more slowly to me; it seemed that I was constructing them at a steady rate. Yet there is nothing that could count for me as a retrospective grasp of such an experience, in its apparent endlessness. (I could not have an apparently endless experience, apparently followed by further experience.) I must, subsequently, have forgotten all but an initial segment of it. How can this be? Surely what we have here is symptomatic of the fact that nothing could ever count, for anyone, as a grasp of an infinite reality. The grammar of 'infinity' is not geared to this. The special problems that arise when we envisage time seeming to go more slowly merely serve to make graphic an incoherence that is there to be acknowledged anyway – an incoherence that crept in at the very beginning of the story. It does not make sense to say that I have just performed infinitely many tasks of any kind, nor to say that anything else is infinite in any respect.[28]

If this is right, then it means that scepticism about infinite co-incidences, and about the law of the excluded middle, such as I advocated in §5, and such as are characteristic of intuitionism, owe as much to the language of infinity as they do to time, however literally or metaphorically understood. And it means that, when we turn to the paradoxes of the infinitely small, we must not be seduced by neat, consistent mathematical formulae into thinking that the relevant super-task stories are really coherent. The formulae need to be *interpreted*.

But there is really nothing new in this. Aristotle took essentially the same line on the paradoxes of the infinitely small some two thousand years ago. This was part of his own view that a completed traversal of the infinite, at any point in time, was impossible: the infinite *was* the untraversable. But now reconsider what was in many respects the most serious difficulty confronting this view, namely the problem posed by infinite past time (see above, Chapter 2, §5). Is there not an analogous problem for the intuitionist? We have been concerned in this section with one objection to the intuitionistic claim that it would be impossible ever to have constructed all the natural numbers. Is there not a second, quite different objection, namely that it would be possible for someone with an infinite history, provided they worked backwards? (Suppose we knew the rate at which they had been working. Then we could say how long ago they started constructing, say, 821, how long they spent constructing all the numbers with a million digits, and so forth.) Is it not a merely physical limitation to have a finite history? (It is certainly not a limitation of temporal immersion, unless time itself is finite. But this, like the possibility that time is quantized, is something that, in this context, we can ignore.)

There is one very obvious intuitionistic counter-objection to this. The construction of a number is based on the separation of time into parts and presupposes the construction of all its predecessors. There is no prospect of constructing the numbers backwards.

But a more basic problem with the proposal is the same as before. Despite the logical consistency of the story, it is fundamentally nonsensical. A hypothesis about somebody's having been around for infinitely long, and having performed infinitely many tasks, is not a hypothesis with which we can *do* anything. There is no provision for recognizing anything as being, essentially, the outcome of an infinite process, nor, therefore, for recognizing any process as being infinitely old. The most that we can say is that certain processes, by their very nature, could never have begun. And similarly, if we do declare past time to be infinite, we can mean no more than this: however long ago any event occurred, other events might have occurred earlier. (Maybe this was all that Aristotle meant.[29]) This relates back to the asymmetry, already noted a couple of times, in our attitudes towards the past and the future: we are more reluctant to think of a process as having no beginning than we are to think of it as having no end (see above, Chapter 2, §5 and Chapter 6, §3). There is no corresponding asymmetry in the past and future themselves. It is just that, because the past leaves traces on the present in a way that the future does not, for example in memory, we are in greater danger, when we declare a process to have no beginning than we are when we declare it to have no end, of committing ourselves to some unacceptable nonsense about an infinite reality that we are *now* confronted with.[30] In *both* cases we must avoid talking in terms of anything's actually being infinite. Such is the grammar of our language.

A final point: it might be objected that the Wittgensteinian critique is too dogmatic. What if we were confronted by a situation where it just did seem appropriate to describe something as infinite?

Well, language is flexible. Grammar can change. We are not making any predictions about our future language use, we are describing it as it is now. If there comes a time when we feel pressure to use language differently (because our current use seems inadequate or inappropriate in some way), then so be it. We must cross that bridge when we come to it.[31]

Meanwhile, even though we must regard it as a special kind of nonsense to say that the truly infinite exists, it remains the case that we are *shown* that it does.

CHAPTER 15

Human Finitude[1]

The pleasure suffusing his body called for darkness. That darkness was pure, perfect, thoughtless, visionless; that darkness was without end, without borders; that darkness was the infinite we each carry within us. (Yes, if you're looking for infinity, just close your eyes!)

... But the larger a man grows in his inner darkness, the more his outer form diminishes. A man with his eyes closed is a wreck of a man. (Milan Kundera)

And the light shineth in darkness; and the darkness comprehended it not. (St John)

My aim in this chapter is to draw together the main ideas that have now emerged, to tie up loose ends, and to combine these into a coherent whole. The focus of the chapter is human finitude, human finitude having been, in its own way, the focus of the whole enquiry.

1 The nature of human finitude

I am metaphysically finite.

What does this mean?

It means that I am cast into a world that is not of my own making. That world does not in any way depend upon me, though I do depend upon it. I depend upon it for my very being. I am precariously balanced on a knife-edge of existence: a minor rearrangement of the particles inside my body would be enough to annihilate me.

It means that I find myself in the midst of (up against, constrained by) something which is *other* than me. There is a reality out there which impinges on me in such a way that I am conscious both of it and of the distinction between myself and what is not myself. (So I can conceive of my never having existed without thereby conceiving of nothing's ever having existed. It is, in a way, remarkable that I can do this.)

It means, or rather it has as one of its most graphic aspects, that I am spatio-temporally (mathematically) finite. I am a particular parcel of matter (consisting mostly of water). That parcel of matter is now alive but will one day be dead. From a metaphysical point of view it is impossible to specify, non-arbitrarily, what my spatio-temporal limits are – when, for example, I began to exist (whether as a foetus, and, if so, at what stage of development), or where I stop and my environment begins (in other words, how big I am). We might think, concerning the second of these, that what counts is the harm that would be inflicted by various abscissions. You can take my shirt off, for example, or even cut my hair off, without harming me much at all. Cutting my thumbs off would be more drastic. Cutting my head off – or perhaps we should say, cutting my torso and limbs off – would be fatal. The distinctive way in which the kind and degree of harm seem to change at a certain point supports the common-sense idea that I am about 1¾ metres tall, weighing between 130 and 140 pounds. On the other hand, if we consider all the rivers of causation in which I am actively involved and then ask what is the (smallest) chunk of the world through which they all flow, or, relatedly, if we ask which chunk of the world is essential to my survival, then I am liable to seem much smaller, say 7 centimetres tall, weighing 3 pounds – sort of greyish, the consistency of yoghurt, and housed inside a skull. This is precisely what some philosophers would say.[2] But we do not need to arbitrate on questions of this kind. The important point is the underlying metaphysical point that there is something definitely beyond me.

I am continually aware of this. Sometimes it is brought home to me in a way that is especially deep and acute. Thus Murdoch:

> 'Falling in love' ... is for many people the most extraordinary and most revealing experience of their lives, whereby the centre of experience is suddenly ripped out of the self, and the dreamy ego is shocked into awareness of an entirely separate reality.[3]

But to be aware of what is outside me in this way, I must be *given* it. This was what Kant took as his philosophical starting point (see above, Chapter 6, §2).[4] But, as he also went on to insist, what I am given must itself be finite. I cannot receive the metaphysically infinite whole. This is because my own finitude sets limits on how much I can take in, and how much I can be affected by what is out there. It is as if my reception is itself a kind of conditioning. If I were going to receive that which was complete, absolute, self-explanatory, and independent of all else, and if I were going to be aware of it as such, I should myself have to become infinite. I should have to be 'absorbed' into the infinite, and lose my (finite) self in it. Hegel believed that this was exactly the kind of thing that was destined to happen. He believed that human thoughts about reality were stages in the development of the infinite's own self-consciousness (see above, Chapter 7, §1). But even on this view, there is never any reception of the infinite whole by any isolable finite part of it.

And *this* is why the concept of the metaphysically infinite whole is incoherent (see above, Chapter 13, §3). If none of us can ever receive anything of this kind, or come to know it, or get our minds around it, or make sense of it, then the concept has no possible application (for us).

It is important to remember here just how much metaphysical infinitude requires. Hawking, writing about the quantum theory of gravity, has said that it

> has opened up a new possibility, in which ... [the] universe would be completely self-contained and not affected by anything outside itself. It would neither be created nor destroyed. It would just BE.[5]

And elsewhere he toys with the idea that there might be no alternative to *how* it would be.[6] But even then it would not be metaphysically infinite in the true sense. As Hawking himself goes on to point out, there would still be a mystery as to *why* it should be. The truly metaphysically infinite would have to quell any such mystery. It would have to explain its very own existence. It would have to offer a theory of itself that was impervious to any future experience, neither open to verification nor vulnerable to falsification.

I could never be given anything like that. Whatever I receive is *ipso facto* set in relation to me. Certain questions therefore inevitably arise about it: for example, how far the way in which it is given to me is a feature of my own particular point of view; or how my reception of it is so much as possible. These questions put it in a kind of focus (alongside me). Even if they have answers, they have answers that arise, so to speak, from outside. It is as if whatever I receive is conditioned by my very ability to receive it. It is finite.

But now we can follow Kant's train of thought (see above, Chapter 6, §2). Not only must whatever I receive be finite, I must be able to recognize it as such. For, as Kant argued, my reception is self-conscious. I must be able to recognize what I receive as *what I receive*. I must be able to reflect on the fact that it is thus and so *from my point of view*. And this in turn means that I must be able to see it in its context, set before me, with, like me, its own conditions of existence. To do this I must be able to recognize, and thus to receive, some of those conditions (whether they be parts of its internal structure, parts of its environment, causes of it, or whatever). So whatever I receive, I must be able to receive more. This sets up a mathematically infinite regress. In some sense, then, the world must present itself to me as mathematically infinite – in the sense, namely, that there must be endless possibilities of further reception written into the form of the finite things that I receive. Another way of putting this is to say that what I receive I must receive in a mathematically infinite framework. (But it is important that this *is* just another way of putting it. I shall return to this point below.) For Kant, space and time were part of such a framework. The main point is that it is a framework of possibilities. Whatever I receive, I must be *able* to receive more.

But the concept of the mathematically infinite cannot be pushed beyond these confines. It has no direct application to reality. In *that* sense, it is every bit as incoherent as the concept of the metaphysically infinite. Indeed it represents an ill-begotten attempt to get *back* to the metaphysically infinite, an attempt to gather together all these possibilities into an actual, self-contained whole. This is not to deny that the concept has a legitimate use. It is just that its legitimate use is not direct. It is, rather, to enable us to characterize all these possibilities. But the possibilities can never all be realized. There is no 'bare filling' of the framework within which I receive things, no infinite co-incidence, no infinite reality. Things afford infinite possibilities. And I can see this. But I see it in their form. The possibilities are never all laid before my gaze in their infinite entirety. This is, just as Wittgenstein insisted, a matter of 'grammar'. There is no provision in our understanding of infinitude for explaining what a direct encounter with the mathematically infinite could be.

In sum, then: my finitude – human finitude – has two fundamental aspects. There is the basic metaphysical finitude of which it is a species. And there is the self-consciousness with which this is overlaid. It is the self-consciousness that gives it its distinctive character. What human finitude *is* is self-conscious metaphysical finitude.

2 Time

We can gain a keener sense of some of these ideas by considering how time fits in. My reception of finite things is as it were indexed. Whatever I receive I receive *at some time*. (I shall not go into the thorny question of how far this is a matter of metaphysical necessity.) What I receive at one time may be different from what I receive at another. This lends substance to the idea that whenever I receive anything it must be possible for me to receive a condition of it. For that possibility may, in fact, be realized – at some other time. (I can probe things, explore them, find out what caused them.) This in turn lends substance to the idea that each finite thing I receive has written into its very form endless possibilities of further reception. Its form can now be understood as including its temporality. We can, with Kant, view time itself as part of the infinite framework of possibilities within which I receive things. Each moment in time essentially points to earlier and later times, just as Aristotle said (see above, Chapter 2, §3). By the same token each reception essentially points to possibilities of earlier and later reception. Recent developments in physics have revealed how time might be finite (both by addition and by division). But even if it is, its structure can still reflect the structure of infinite possibility in this way. And this helps to explain why, ever since the time of Aristotle and indeed before that, an intimate association has always been recognized between time and the infinite. On the intuitive (perhaps mistaken) understanding of time as never-ending, this association is more intimate still, and helps to reinforce the *bona fide* conception of the mathematically infinite: each time is necessarily succeeded by others, but there is no such thing as the whole of time, conceived as an 'infinite reality'.

For anyone of a vaguely Hegelian bent there is scope for recognizing even deeper connections. The incoherence of the metaphysically infinite whole can be seen as broken apart into finite, manageable, coherent parts, each capable of being given at some particular point in time, in such a way that, over time, there can be – how to put it? – a progression back towards the whole. Think how in set theory, for example, there is no Set of all Sets: but, in the metaphor that we have been deploying, there is, over time, a steady accumulation of Sets, each appearing sooner or later.

3 The infinite as an Idea of reason. The saying/showing distinction revisited

We have a concept of the metaphysically infinite. But it has no possible direct application to reality.

We also have a concept of the mathematically infinite. This enables us to characterize the endless possibilities that things afford. I can say, for example, that I receive things in a mathematically infinite framework. But this does not mean that there is an infinite framework that is filled with the things I receive. In fact there is no such *thing* as the framework at all. (Certainly it is not itself another thing that I receive.) For nothing at all is mathematically infinite. In this respect our concept of the mathematically infinite is just like our concept of the metaphysically infinite: it has no possible direct application to reality.

But the connection cuts deeper than that. To try to apply the concept of the mathematically infinite directly to reality would be, as I intimated in §1, to try to get back to the metaphysically infinite – back to the unity of the whole. The connection can thus be seen in broadly Kantian terms. That is, there are not two separate concepts. There is a single concept, the concept of the infinite, capable of being viewed under two different aspects (as it were, the aspect of how things are in themselves and the aspect of how we receive them: these correspond to the metaphysically infinite and the mathematically infinite respectively). Any *tension* between the two simply reflects the incoherence in the (one) concept.

Still, we do *have* this concept. And it is clearly not read off from experience. (This was what created problems for the empiricists.) How then are we to view it? We can, I think, take up a suggestion which has recurred many times throughout this enquiry and which was endorsed, for example, by Hilbert (see above, Chapter 9, §2). I mean the suggestion that our concept of the infinite is a Kantian Idea of reason, or at least that it is something like a Kantian Idea of reason. (The qualification is important: Kant allowed for the direct application of Ideas beyond the range of our knowledge and experience, but not even this concession is being made here.) It is a concept that we cannot apply directly. (When we try to do so, 'human reason precipitates itself into darkness and contradictions.'[7]) But it does have a regulative use. Indeed that *is* its legitimate use. To talk about the 'infinite framework' in which things are received, for example, is, in effect, to say, 'Proceed *as if* there were an infinite reality out there.' And this provides a continual spur to realize the possibilities inherent in things. It provides a continual spur to further reception. It leads, ultimately, to an increase in knowledge and understanding.

But what first induces us to talk in this way? What induces us to say (to ourselves) that we should proceed as if the infinite existed?

Answer: the fact that there is an urge, which has to be tamed, to say that the infinite does exist (an urge, in other words, to apply the concept of the infinite directly).

But where does this urge come from? There is a contrast with our own finitude: but why is there a temptation to suppose that anything stands on the other side of the contrast?

This is where the saying/showing distinction comes in. We have an inexpressible insight, and the urge in question is none other than the urge to express it. I shall now try to offer further elucidation and defence of this contention.

Again the focus is Kant. More specifically, the focus is the section in his *Critique of Pure Reason* entitled 'The Deduction of the Pure Concepts of Understanding'.[8] This passage was the source for the discussion of self-consciousness in §1. In it Kant argued that self-conscious reception, such as mine, or, if you like, self-conscious awareness of the world, consisted of a variety of elements that needed to be held together in unity. There needed to be a one-ness in their many-ness. In Kant's own terms there needed to be a 'synthetic unity of the manifold'.[9] (See above, Chapter 14, §3.) This is surely right. Unless there were such unity, I could not self-consciously recognize all the things I receive *as* things I receive. But this unity is both a unity in my experiences (a unity in the *receiver*) and a unity in things in the world (a unity in the *received*). These are two sides of the same coin. In neutral terms, or rather in terms that allude at once to both sides, it is a unity in my experiences of things in the world (a unity in the *reception*). To adapt a phrase used by Wittgenstein in the *Tractatus*, it is a unity in the world as I find it.[10] Self-conscious introspection affords a kind of access to this unity. But, as Kant insisted, it is not itself something I receive. It is more like a condition of my receiving anything at all. So it is not itself a thing *in* the world, like other things. As a result, I cannot talk about it, in any straightforward way. If I attempt to do so, I am liable to produce nonsense of one of two kinds, corresponding to the two aspects of the unity. On the one hand I might start talking about myself, as an isolable substance: here I would be trying to talk about the unity in the *receiver*. (For reservations about the idea that I am an isolable substance, or that I am clearly delineated from my environment, see above, §1. Kant railed against such talk in a later section of his book.[11] It received its clearest expression in Descartes' celebrated distinction between mind and matter.) On the other hand I might start talking about the world as an infinite whole: here I would be trying to talk about the unity in the *received*. These two kinds of nonsense are the obverse of each other.[12]

The second kind of nonsense involves me in using the Idea of the infinite non-regulatively. But once my urge to do this has been curbed, I can use the Idea regulatively instead. I can enjoin myself to proceed *as if* the unity were given to me, as something lying open to investigation. And this, as we saw, can lead to an increase in knowledge and understanding. It is *this* fact that entitles us to talk about an insight here. My urge springs from a genuine state of enlightenment. It is just that it is an inexpressible state of enlightenment. I am shown something. What I am shown can be specified in terms of the corresponding urge (see above, Chapter 13, §2). I am shown that the world exists as an infinite whole. I am shown that the infinite exists.

In fact, however, whatever exists is finite. So in a sense, I am shown the finitude of infinity. I am shown that the infinite world exists as a limited, finite whole. (This reflects the incoherence in trying to apply the Idea of the infinite directly.) The finitude of the whole is a matter of all its parts being bound together in unity. This unity is the unity in my reception. As such, it is rooted in my own finitude (precisely what determines that anything I receive must be finite). So I am, in effect, shown everything bound together by my own finitude. I am shown, to quote the *Tractatus* once again, that 'the world is my world; the world and life are one; I am my world.'[13] (I am shown, despite what was said above in §1, that if I had never existed, then nothing would have.) To be shown these things is part and parcel of what it is to be metaphysically finite in the way that human beings are, that is to say, self-consciously so. Recall what was said above in Chapter 13, §2: to know something that can be put into words is to receive; to be shown something is to be self-consciously receptive.

What else am I shown? Not just that the infinite exists but that fundamentally *only* the infinite exists. I said in Chapter 10, §5 that points are where lines do such things as stop; and lines are where surfaces do such things as stop; and surfaces are where bodies do such things as stop. This is reminiscent of an idea that has come to the fore in this chapter: that each reception points to possibilities of further reception. (If I attend to a surface, say, then I am immediately presented with the possibility of attending to the body whose surface it is.) More strikingly, each finite thing that I receive points to the whole infinite framework of possibilities in which I receive things.[14] This leads (or so I shall suggest – the inference is obviously not incontrovertible) to the idea that, if there is an infinite whole, then everything is merely an aspect of it, 'where it does something'. But that everything *is* merely an aspect of the infinite whole is not something that I can be said to know. It is something that I am shown.

Something else that I am shown is that a certain kind of relativism holds – relativism of the kind that was discussed in connection with the Löwenheim-Skolem theorem (see above, Chapter 11, §§2–3). For I am shown that all the things I receive are bound together into a whole; and this means not just that the whole is finite but that (because it is finite) it excludes certain other things, in other words there are things that I do not receive – though they may be received 'elsewhere'. Hence the relativism. Compare: I am shown that what I understand to be the full hierarchy of Sets is just another Set; so from elsewhere there is access to further Sets.[15]

But we must not forget that my being shown these things in no way entails their truth, or even their coherence. There is not in fact a Set of all Sets. Nor does the infinite exist.

This is an apt point at which to return to the main paradox of thought about the infinite, whose solution, I said, would be a key to the whole enquiry (see above, Introduction, §4). The paradox is that I appear to be able to grasp the infinite as that which is ungraspable. The solution is to deny that I can grasp it in any way. There is nothing there to be grasped. It is just that self-conscious introspection leads me to an insight which I have an urge to express by applying my Idea of the infinite directly to reality. But I must not do this. The insight is in fact inexpressible.

Postscript and preview

I tried to make clear in Chapter 13 how far these ideas have their source in Wittgenstein. The solipsism, in particular, has been lifted straight from the *Tractatus*. But it was there already in Kant. This marriage of their systems is not a simple eclecticism. In the first edition of the *Critique of Pure Reason* Kant wrote:

> All objects with which we can occupy ourselves, are one and all in me, that is, are determinations of my identical self . . . [This] is only another way of saying that there must be a complete unity of them in one and the same apperception. But this unity of possible consciousness also constitutes the form of all knowledge of objects; through it the manifold is thought as belonging to a single object.[16]

This highlights a crucial line of influence from Kant to Wittgenstein. And this in turn is a good cue for mentioning a philosopher who, though he did not appear in Part One of this book, must now, because of how the enquiry has panned out, be given his due for having played a vital role in the whole drama: the German Arthur Schopenhauer (1788–1860). The line of influence from Kant to Wittgenstein passed directly through him. In his own work Schopenhauer owed a great deal to Kant. Wittgenstein, for his part, when he wrote the *Tractatus,* was deeply indebted to Schopenhauer.[17] An enormous amount was propagated in this way. In particular this is true of the solipsism.

But Wittgenstein was indebted to Schopenhauer in more specific ways. For example, the analogy of the field of vision (see above, Chapter 13, §1) derived from Schopenhauer.[18] So too did many of Wittgenstein's views on death and immortality (see above, Chapter 13, §3). By focusing on what I am shown, and viewing my world as a limited whole (whose value attaches to it *as* a whole) I can cease to fear death as the end of anything or as robbing me of anything; there is nothing beyond my life, or beyond my world. (That is, there is nothing *for me*. This qualification picks up on the relativism just discussed.) This Wittgensteinian thought had been eloquently anticipated by Schopenhauer as follows:

> Just as on [a] globe everywhere is above, so the form of all life is the *present*; and to fear death because it robs us of the present is no wiser than to fear that we can slip down from the round globe on the top of which we are now fortunately standing.[19]

Wittgenstein wrote, in a similar vein, that 'eternal life belongs to those who live in the present'.[20] Of course, mortality does lend poignancy to my life. It does make my finitude that much more graphic. But these are not a matter of my (so to speak) one day being confronted with the fact of my own non-existence, nor exactly of my being deprived of something that would be there for me to enjoy if only I were immortal. In the remaining two sections of this chapter I shall try to develop these themes.

4 The poignancy of human finitude. Death[21]

My finitude can, at times, give me a sense of abandonment, isolation, absurdity. I can feel forsaken. I can feel cast adrift in an alien environment that constrains me in various ways, thwarts me in various ways, bewilders me, scares me, and grieves me. And my temporal finitude can exacerbate this. There was a time when I did not exist: I was placed in the world. And there will come a time when I shall no longer exist: I shall be taken from the world. (This begs the question of whether I have an afterlife. I beg it with reluctance. But I think that the possibility is too ill-conceived for there to be any advantage in trying to accommodate it.) These facts, in particular the fact that I shall die,[22] can, all too easily, induce in me a special sense of pointlessness.

But why? As Wittgenstein asked in the *Tractatus*, would any riddle be solved by my surviving for ever?[23] Surely there is no kind of meaning that could attach to my life if I were immortal that could not attach to it anyway. It is not even obvious that, if I were immortal, my life would have *more* meaning. Indeed it might have less. For my mortality does lend shape to my life. It lends shape to every project I undertake and every scheme I embark upon.[24] Far from its being the aspect of my finitude that most fundamentally threatens me with a sense of absurdity, it might actually help to alleviate it.

Of course, even if I admit this, I may still never want to die. Consider the following quotation from Nagel:

> Given the simple choice between living for another week and dying in five minutes I would always choose to live for another week ... I conclude that I would be glad to live forever.[25]

That does not follow. I might be appalled at the thought that I shall live for ever, without, at any particular time in the future, wanting *these* to be my last five minutes. (That is, I might never want to die without wanting never to die.) More starkly, I might be appalled at the thought that I shall live for ever *and* appalled at the thought that I shall one day die. Or, rather differently – we shall see soon why this is different – I might be appalled at the prospect of infinite conscious existence (that is, conscious existence which, for some reason, can never end) *and* appalled at the prospect of finite conscious existence. There is no reason why either option should attract me – though there is no third alternative.

Let us use the term 'nothingness' to describe a period of time in which, except perhaps for the ticking of a natural 'clock' somewhere in the universe, there is complete inactivity – a period during which no sentient being has any experiences, and immediately after which everything returns to precisely the state that it was in before. Then not even long periods of nothingness could help to make either alternative more attractive. Such periods would neither relieve the tedium of infinite consciousness, nor usefully put off the impending *dénouement* of finite consciousness. For by their very nature they could have no direct effect on me. If a trillion years of nothingness were to begin in five minutes' time, I should be none the wiser. Once they had elapsed, it would seem to me as if everything had carried on continuously. The plain fact is that between infinite conscious existence and finite conscious existence there is no middle ground.

I want to suggest, in the light of these reflections, that my death is not the crux of my finitude. There is not some kind of *sense* or *solidity* that would attach to my life if only I were immortal. (It is of course true that my dying at any *particular* time will cut me off from certain possibilities. That is precisely why I may never want to die. I shall return to this point later.) These are truths that I shall find that much easier to grasp if I am shown that my death does not rob me of anything, and that it is not even the end of anything. Conversely, the more the differences between mortality and immortality are played down, the easier it will be for me to be shown these things. Let us consider some particular ways of playing down the differences that serve this purpose very well.

A story was told in Chapter 14, §6, in which I performed infinitely many tasks in a minute and, while I was doing this, time seemed to be going more and more slowly to me, so that the whole operation seemed endless. The point was actually to locate an incoherence in the story. But we are reminded that periods of time can seem longer or shorter than they really are. Could something similar happen 'in reverse'? That is, could time seem to go more and more quickly to me (as in fact it does) in such a way that an endless period seemed to last for only forty years, say? (The first twenty years might seem like twenty years, the next twenty like ten, the next twenty like five, and so on *ad infinitum*.) The suggestion strikes us as puzzling because we are left wondering how it could ever seem to me that the forty years were up. But of course, given a real period of forty years, at the end of which I *died*, it would not in fact ever seem to me that they were up. So might it not be like that?

Here is a story that makes a similar point.

The story of my veiled immortality

Time seems to be going neither more and more slowly to me nor more and more quickly. I live normally for twenty years, and these are followed by a trillion years of nothingness. I then live normally for a further ten years, and these in turn are followed by a trillion years of nothingness. I then live normally for five years, which are in turn followed by a trillion years of nothingness. And so on *ad infinitum*. I never die. But it is (to me) *as if* I live for just forty years.

Comment: The important thing about the trillion-year bouts of nothingness is that they are all the same length. The story could just as well have been told, curiously enough, with intervals of a trillionth of a second.

This story demonstrates how a certain kind of immortality could seem to me just like mortality. (And this would not just be in the following, boring way: I could, after a certain time, live for ever as a vegetable with no conscious existence. In the story above, no matter how far ahead in the future you consider, there are periods of *normal life* left for me.) The striking thing is, if such immortality would seem to me like mortality, then, by symmetry, my mortality must seem to me like a kind of immortality. It is (to me) *as if*, in some strange scenario, I shall always be alive, and indeed as if I always was alive (for a similar story could be told about my past). The point is, of course, that there never comes a time when I am able to reflect, 'That is it. The forty years (or whatever it may be) are now up. I am dead.' My death never *comes for me*. To quote Wittgenstein again:

Death is not an event in life: we do not live to experience death.[26]

And, as I contemplate this fact, it is brought home to me that I receive nothing after my death. This helps me to reflect self-consciously on the nature of my reception. I am then shown that there *is* nothing (for me) after my death; my world is a limited whole, and value attaches to it *as* a whole. These are things that I am shown in much the way that I would be if I were immortal. The difference between mortality and immortality is not the key here.

It is not, of course, that my death is unimportant. Indeed much of its importance lies in the fact that what I am shown is false: death does rob me of something. But this is a matter of its occurring *when* it does, not of its occurring at all (that is, not of my being mortal). And in this respect it is nothing but a graphic illustration of something deeper, something closer to the essence of my metaphysical finitude: *the closing off of possibilities*.

I argued in §1 that, because of my metaphysical finitude, I am confronted with limitless possibilities. This confrontation is temporal. The possibilities that I am confronted with at one time may, at another time, be realized. Equally they may not. They may remain 'mere', unrealized possibilities. More pertinently, they may become 'mere', retrospective possibilities – possibilities which might have been realized but which now, for one reason or another, can no longer be. For example, if I come into contact with something that is subsequently and irrevocably destroyed, then I am no longer able to probe it or explore its conditions. Again, if I become incapacitated in some way, then I am prevented from realizing some of the possibilities that were once open to me. It is in this sense that we can talk of the 'closing off of possibilities' (just as, when I acquire new capacities, we can talk of the 'opening up of possibilities'). And of all the things that can close off possibilities for me, my death is the most obvious and starkest example.

In fact it can play a part in doing this even before I die. For the sheer fact that I shall die, within a certain period, means that, as I grow older, various things cease to be possible for me: I no longer have the time for them. (One very important consequence of this is that there are certain possibilities which I cannot realize except at the expense of others.) But there is a sense in which my death simply highlights a fundamental feature of my metaphysical finitude that would have been there even had I been immortal. I am cast into a temporal world whose constraints and vicissitudes mean that, as I journey through it, I continually see alternative routes being blocked off.

Of course, Heidegger was right to insist that my death has a unique importance as a parameter for all that matters to me. But the more direct emphasis of other existentialists on the play of possibilities may have got closer to the kernel of my finitude. My death is certainly not alone in closing off possibilities for me. I can become incapacitated in much less extreme ways. Somebody else, who is close to me, can die. A relationship can die. Things can get forgotten, overlooked, passed over, rejected, renounced, or lost.

And make no mistake: the closing off of possibilities is, quintessentially, harmful.[27] (This is what gives human finitude much of its poignancy.) In particular – I make no apologies for the banality of this – death is harmful (which is not to deny that somebody may, for quite undramatic reasons, be better off dead).[28] Nothing that I have said in this section is meant to belittle the special – indeed absolute – destructiveness that pertains to death. In particular, my dying *when* I do may well deprive me of much that would have enhanced my life had I lived longer. (Unjustified killing is, in its finality and absoluteness, uniquely evil. It is a terrible irony that technological advances have opened up new possibilities of killing, so that human beings are now in a position, at pretty much the flick of a switch, to close off *all* their possibilities.[29]) Whenever possibilities are closed off for me, parts of 'the other' escape me in its infinite and inexhaustible variety. I can no longer receive them. The possibilities can no longer beckon me or guide me or give sense to my life, as they could while they still confronted me. When possibilities confront me, they are possibilities *for me*. When they are closed off, I lose something of myself. Sometimes I can accept this with equanimity. Sometimes it is unbearably painful.

5 Being finite

The basic premise for this chapter has been that I, like all human beings, am (metaphysically) finite. Kant, along with many others, though with greater boldness and with a purer vision than most, insisted that there was also, deep within me, essential to my very being, a metaphysical infinitude: the metaphysical infinitude of reason. If this were right, the argument in §1 for the incoherence of the concept of the metaphysically infinite would collapse. The metaphysically infinite would be something that I could 'see before me and associate directly with the consciousness of my own existence'.[30] Doubts about any direct application of the concept of the mathematically infinite would have to be surrendered too. And in fact Kant did argue that I could only make sense of my rationality in mathematically infinite terms. I had to believe that I was immortal, not just in the sense that, however long I lived, I should always have longer to live, but in the more robust sense that my life could be viewed as an infinite progression to full atonement (see above, Chapter 6, §4).

Kant's vision was an extraordinary and beautiful one. But its accompanying metaphysical baggage proved too heavy, and it eventually involved him in incoherence (see above, Chapter 14, §2). Once that baggage has been discarded, there is no clear sense in which my rationality, or my potential for goodness, or indeed anything else, make me in any respect metaphysically infinite. I am finite, and must come to terms with my finitude. Even when I am shown my life as a metaphysically infinite whole, this merely serves to emphasize my finitude. What I am doing is reflecting, self-consciously, on how I, as a finite being, receive 'the other'. My finitude is not something I can escape. All I can do is to (try to) master the art of being finite.

It is way beyond the scope of this book to provide a useful elaboration of this idea. But I shall close with some sketchy remarks about it.

I am shown that my life is metaphysically infinite: I am shown that my life, my world, I, form a self-contained whole. But what would it be for me to sustain this insight and, as it were, to live by it? (For reasons that will emerge, I do not want to suggest that it would be easy for me to do this. It might not even be possible.) I should have to recognize all the finite things that exist as bound together in a kind of unity. But there is a paradox. For, despite what I am shown, the finite things that exist include me and my actions. These too are merely a part of the world, bound together with everything else. I should have to accept that. (For example, I should have to overcome any feelings of alienation that I have in the face of 'the other'; likewise any sense of abandonment induced by my finitude (see above, §4).) This is one of the hardest things for me to do. In fact the greatest temptation that I face is the temptation to set myself apart. I crave self-sufficiency and independence. I crave metaphysical infinitude. I should have to overcome that. This is why the paradox is so acute: it is only then that I could be fully shown myself as – precisely – metaphysically infinite.[31]

The Kantian vision provides one powerful and important attempt to give voice to this. Kant held that for me to act rationally was for me to act on a maxim which I could accept as a universal law (in effect, and very roughly, to do as I would be done by). But this is how I would act if I really did regard myself as just another part of the world – if I were shown that everything, myself and my actions included, were bound together in a harmonious whole. So if we wanted to see this in Kantian terms, we could say that I was being shown the unifying power of reason; I was being shown reason itself, in its all-embracing, impartial, metaphysical infinitude. Moreover, I was being shown it as something that lay deep within me and determined the course of my actions.

In a more Wittgensteinian vein we could say that I was becoming one with the world, or that I was in agreement with 'the other'.[32] 'The other', of course, includes other people. They too would be integrated into the whole that I was shown. I should have to recognize that they too were bound up with me, and that their call on me was every bit as urgent as my call on them.

A further crucial aspect of my living by what I was shown would be an acceptance of what was possible and what was not. We have seen how this can hinge on my finitude. And there is nothing I can do about that. There is nothing I can do about the fact that I am lodged here rather than there, now rather than then: I am confronted with these possibilities rather than those. To be in agreement with 'the other' – to accept my particular standing in the world – I should have to assimilate these possibilities and become alive to them. It is seductively easy to brood over closed-off possibilities instead and to dwell on what might have been. But this distracts, and it destroys. Its net effect is that yet more possibilities get closed off. (They get ignored.) I should have to resist that. I should have to be shown my world expanding to accommodate the possibilities that were *now* open to me.[33]

But how much of an expansion could this be? Enough to give my life any real sense?

Here, I suggest, there is room for *hope*. St Paul said that there were three things that lasted for ever: faith and love were two, the third was hope.[34] Kant said that all my rational interests were combined in the following three questions:

> What can I know?
> What ought I to do?
> What may I hope?[35]

For both St Paul and Kant, hope had its place at the very centre of my moral experience. But what place can it have here?

Without it, I cannot have any confidence that anything has any point (any more); I cannot properly live by what I am shown. My hope is twofold: firstly, that, whatever happens, the most important possibilities will never be closed off; and secondly, concerning my own life, that, for as long as I am confronted with any possibilities at all, I shall be confronted with possibilities that give it sense. In hoping these things, if I do hope them, I am using my Idea of the infinite in a regulative way. I am proceeding *as if* what ultimately mattered enjoyed a kind of infinite resilience.

But how can I hope these things? How can I properly live by what I am shown?

It may be that I cannot. Or it may be that I cannot, unaided. These are questions that have to be *lived*. They are not questions to be tackled here.

There are various related questions about my temporality which must also be left unanswered. Living by what I am shown means living in the here and now (assimilating the possibilities that confront me here and now) yet with an eye to eternity (the eternity of my life as it is shown to me). In Wittgenstein's words, it means living 'not in time but in the present'.[36] But, as regards my past, how far is this a matter of turning my back on it, and how far is it a matter of drawing on what is available here and now to make *sense* of it? Reconsider the debate between Kant and Nietzsche (see above,

Chapter 7, §5). If it is always possible for me to make sense of my life, is this because, no matter what I have done in the past, there are specific possibilities, including possibilities of repentance and atonement, that will always lie open to me, as Kant believed? Or is it because, whatever things I have done in the past, they can always be creatively woven into the continuing saga of my life, as Nietzsche believed? In making sense of my life I must be shown it as a whole. This seems to support the Nietzschean alternative. On the other hand, turning my back on the past is not the same as ignoring it. To repent is not to forget.

I have finished on an inconclusive note. But it is significant that these questions have so much as arisen. For it serves to emphasize something that I think has been apparent throughout the enquiry, particularly in these last few Kant-imbued sections: how far the problems of philosophy are problems, ultimately, about human finitude. Given that problems about human finitude are themselves problems about the infinite, this in turn helps to explain why any enquiry into the infinite, insofar as it can be taken seriously, is bound to cut deep. It must raise questions of the most fundamental kind about the world, about us, and about our place in the world. It must cut to the very heart of philosophy.

Glossary

$\aleph_\alpha$: the infinite cardinal whose index is the ordinal α (Chapter 10, §3)

analysis: the theory of the real numbers (Chapter 4, §1). (The adjective corresponding to 'analysis' is 'analytic'.)

apeiron, to (Greek): the unlimited or unbounded, the infinite (Chapter 1, §1)

arithmetic: the theory of the natural numbers (Chapter 1, §5; see also Chapter 12, §2)

cardinal number (or cardinal): number which registers the size of a set (Chapter 10, §3). (An infinite cardinal registers the size of an infinite set.)

continuum hypothesis, the: Cantor's hypothesis that $2^\circ = \aleph_1$ (Chapter 10, §3). (This is equivalent to the hypothesis that the power-set of $\mathbb{N}$, or $\mathbb{R}$, is the next infinite size up from $\mathbb{N}$.)

countable (of a set): either finite or the smallest infinite size (Chapter 10, §3)

$\mathbb{N}$: the set of natural numbers (Chapter 8, §3)

natural number: non-negative whole number (Introduction, §2). (The natural numbers are 0, 1, 2, . . .)

Ω: the set of ordinals (Chapter 8, §4)

ω: the first ordinal to succeed all the natural numbers (Chapter 8, §4)

ordinal number (or ordinal): number which registers the length, or shape, of a well-ordering (Chapter 8, §4). (The ordinals are 0, 1, 2, . . ., ω, $\omega + 1$, . . .)

peras (Greek): limit or bound (Chapter 1, §1)

power-set: set of subsets of a given set (Chapter 8, §3)

Glossary

ℝ: the set of reals (Chapter 8, §3)

rational number (or rational): quotient of two whole numbers, negative or
  non-negative (introduction, §2). (The rationals are numbers of
  the form p/q, where p and q are whole numbers and q is not 0.)

real number (or real): number which can be expressed using an infinite
  decimal expansion (Chapter 1, §2; see also Chapter 4, §3)

Set (capital 'S'): set which would still exist even if there were nothing *but*
  sets (Chapter 10, §1)

uncountable (of a set): not countable (Chapter 10, §3)

well-ordering: imposition of order on the members of a given set X which
  singles out, for each set of members of X that have already been
  singled out, a first to succeed them all, unless there are none left
  (Chapter 8, §4)

ZF: Zermelo-Frænkel set theory (Chapter 8, §6)

Notes

Introduction

1 Cf. J. Thomson (1967).
2 *Physics*, III, 4, 204a in Aristotle (1984).
3 *Physics*, VI, 9 in Aristotle (1984).
4 Benardete, pp. 259–60.
5 See Sorabji, p. 218.
6 Benardete, pp. 113ff.
7 Benardete, Ch. VI, in the discussion of the 'serrated continuum'.
8 Respectively: Cantor (1932), p. 282 (also Cantor (1955), p. 85); and Cantor (1932), p. 204.
9 Cf. Enderton (1977), pp. 1–2.
10 See below, Ch. 9, §2; and see Ch. 9, notes 6 and 7 for references.
11 Russell (1946) and (1978) provide helpful background, and I am indebted at various points to both.

1 Early Greek Thought

1 Helpful background reading for §§1–3 is Barnes (1979). The main source is Barnes (ed.) (1987). Most of the material for these three sections is derived from Moore (1989).
2 Barnes (ed.) (1987), p. 75. See further for this section Barnes (ed.) (1987), pp. 71–6; Kerferd; and Seligman.
3 See *Physics*, III, 4, 203a1–8 and IV, 6, 213b22–7 in Aristotle (1984).
4 The Greeks used the word '*logos*'. This is variously translated as 'reason', 'word', 'measure', and 'principle'.
5 See further Barnes (ed.) (1987), pp. 81–8 and 202–13; and *Metaphysics*, I, V in Aristotle (1984).
6 These are the titles of the two parts of his poem 'On Nature' in Barnes (ed.) (1987).
7 Barnes (ed.) (1987), p. 135. See also Furley. I have adopted a translation that partly follows Barnes' and partly follows Furley's.
8 See Barnes (ed.) (1987), pp. 96–7.
9 I do not pretend that this is unproblematical. I shall return to the problems in Part Two (see esp. Ch. 13, §3).
10 Barnes (ed.) (1987), pp. 143–9.
11 As I mentioned in the introduction, we do not know how Zeno himself cast them. All four are presented in *Physics*, VI, 9 in Aristotle (1984).
12 *Physics*, VIII, 8, 263a in Aristotle (1984).

236

Notes

13 Cf. Russell (1937), §322. This essentially involves the contrapositive of Aristotle's proof as set out in Ch. 2, §4.
14 See Simplicius, 140, 28D.
15 See further Vlastos. For references to work on Zeno's paradoxes see Ch. 4, note 16.
16 *Timaeus*, 53ff. and *Parmenides*, 158b–d and 164c–5c in Plato.
17 *Philebus*, 23–31 in Plato.
18 *Republic*, esp. V–VII, in Plato.
19 *Phaedo* and *Meno, passim* in Plato.
20 *Timaeus*, 37d–8b in Plato.
21 *Timaeus*, 33 and 44d in Plato.
22 *Philebus*, 15b in Plato.
23 E.g. *Parmenides*, 132a–3a in Plato.
24 See further for this section Sayre, esp Ch. Three.
25 Helpful background reading for this section are Heath; and Lasserre.
26 *Physics*, III, 7, 207b27–34 in Aristotle (1984).
27 This claim is admittedly open to challenge, given modern standards of rigour. See Lear (1988), pp. 215–18. Cf. below, Ch. 8, note 48, and Ch. 12, note 1.
28 Euclid.
29 See further Boyer, Ch. II; Kline, Chs. 3 and 4; Knorr; and Priestley, *passim.*, esp. Ch. 2.

2 Aristotle

1 Helpful background reading for this chapter are Lear (1988); and Ross.
2 'Empiricist' is a term that Kant was later to ascribe to him. See Kant (1933), A854/B882.
3 See Barnes (ed.) (1987), p. 229.
4 See Introduction, note 2.
5 *Physics*, III, 6, 206b27 in Aristotle (1984). Plato's *apeiron* (the indeterminate) had allowed for determinations of two kinds: beyond ever more remote extremes; and within ever narrower limits. And this may have been what Aristotle was alluding to. But see also *Parmenides*, 164c–5b in Plato. A general reference for this section is *Physics*, III, 4 in Aristotle (1984).
6 General references for this section are *Physics*, III, 4 and 5, and *De Caelo*, I, 5–7 and 9 in Aristotle (1984).
7 *Physics*, VIII, 1, 251b19–29 in Aristotle (1984).
8 There is some exegetical debate about this. See Hintikka (1966); and Lear (1979–80).
9 E.g. *De Caelo*, I, 12 in Aristotle (1984). See also Hintikka (1957).
10 See Lear (1982).
11 *Physics*, VI, 2 in Aristotle (1984).
12 *Physics*, III, 7, 207b21–7 in Aristotle (1984).
13 *De Generatio et Corruptione*, I, 2, 316a–7a in Aristotle (1984).
14 Aristotle (1983), III, 6, 206a27–9.
15 Aristotle (1983), III, 6, 207a6–8.
16 Aristotle (1983), III, 6, 206b34–7a1.
17 General references for §§3 and 4 are *Physics* III, 6–8; VI, 9; and VIII, 8 in Aristotle (1984).
18 See Bennett (1971), p. 135. Cf. Wittgenstein (1975), p. 166.
19 See Bennett (1971); Bennett (1974), Ch. 7; and Waismann.
20 *Physics*, III, 6, 206b1–3 in Aristotle (1984).
21 *Physics*, VIII, 1, 251b10–28 in Aristotle (1984).

22 See Lear (1979–80). See further Sorabji, Ch. 14. For further discussion, see below, Ch. 14, §6.

3 Medieval and Renaissance Thought

1 Helpful background reading for this chapter are Duhem (1985a); Duhem (1985b); Koyré; Kretzmann (ed.); and Sorabji.
2 Plotinus, VI, 9, 6.
3 Plotinus, VI, 6, 8.
4 Plotinus, V, 5, 11.
5 Plotinus, III, 7, 11. A general reference for the above is Plotinus, V and VI, *passim*.
6 Augustine (1945), XII, 18.
7 Augustine (1961), XI, 11–14.
8 Augustine (1945), XII, 20.
9 Augustine (1945), XI, 10–13.
10 See Sorabji, Ch. 14.
11 Still, this is a curious reversal of modern intuitionistic orthodoxy, where some principle of generation and ordering is precisely what *prevents* an infinite set from being unproblematical (see below, Ch. 9, §1). See further Dummett (1977), pp. 55–6.
12 See Duhem (1985a), I, *passim*; and Sorabji, pp. 225–6.
13 Aquinas, I, vii, 1.
14 Aquinas, I, ii, 3.
15 Aquinas, I, vii.
16 Aquinas, I, xlvi, 2.
17 Helpful background reading for this section, with references, are Duhem (1985a); J. Murdoch (1982a); and J. Murdoch (1982b). I am especially indebted to Duhem (1985a), I.
18 Galileo, pp. 21–40.
19 Nicholas of Cusa, I and II, I.
20 Bruno. See also Koyré, esp. Ch. II; and Lovejoy, pp. 116–21.

4 The Calculus

1 Helpful background reading for this chapter are Boyer; Grattan-Guiness (ed.) (1980a); and Priestley.
2 See further Lang, esp. Chs. II, III and IX; and Priestley, *passim*.
3 See Pedersen.
4 Newton (1779–85), I, p. 333.
5 Newton (1779–85), I, p. 25.
6 General references for the above are Leibniz (1962); Newton (1946); Newton (1736); and Newton (1744). See further Bos; Boyer, esp. Ch. V; Kline, esp. Ch. 17; and A. Robinson.
7 In Berkeley (1901). The famous quotation at the beginning of this chapter is taken from this essay, §35. The 'infidel mathematician' to whom it was addressed was Halley.
8 De l'Hôpital.
9 Euler.
10 Hegel (1969), pp. 238–314.
11 The symbol '∞' is often used to denote the infinite. It was first used in this way in the seventeenth century by Wallis.

12 See below, Ch. 9, §3 for criticism of this way of putting it, heralding from Wittgenstein.
13 'Continuity and Irrational Numbers' in Dedekind.
14 For exposition of non-standard analysis see Enderton (1972), §2.8. See further for the material in this section Boyer, esp. Chs VI and VII; Grattan-Guiness (1980b); Kline, esp. Chs 19, 24 and 30; Priestley, *passim*; Quine (1960), §51; A. Robinson; and Tiles, Ch. 4.
15 J. Thomson (1954); and Benardete, p. 149.
16 There is a vast literature on these issues. See further e.g. Benacerraf (1962); Beresford; Black; Bostock; Chihara; Grünbaum (1968a); Grünbaum (1968b); Grünbaum (1973); Lear (1981); Moore (1989–90); G. Owen; Ryle, Ch. III; and Salmon (ed.). (Beresford argues that it is physically necessary for the lamp to be on at the end of the minute. But since we are already in the realms of the physically impossible, I think the argument loses much of its force.) For further discussion see below, Ch. 14, §6.
17 This is anticipation of themes that will dominate Part Two.

5 The Rationalists and the Empiricists

1 *Principle* 26 and *Third Meditation* in Descartes (1984).
2 Pascal, §§199–202.
3 *Fifth Set of Objections* in Descartes (1984).
4 Descartes (1964–75), V, p. 356, my translation. See also *Objections and Replies, passim* and *Principle* 27 in Descartes (1984).
5 Spinoza, I, Defs. 2, 6 and 8 and Props. 8, 11, 13, 15, and 21–6. See further Bennett (1984), *passim*.
6 Spinoza, I, Defs. 2 and 6.
7 Leibniz (1960–1), I, p. 416.
8 Leibniz (1981), p. 57.
9 Leibniz and Clarke. But note that Newton himself denied that the physical world could be finite. See Hawking, p. 5; and Maor, pp. 202–3.
10 Leibniz (1960–1), III, p. 622.
11 Leibniz (1960–1), I, p. 338.
12 General references for the above are Leibniz (1981), II, 17 and 29, 15–16. See further Ishiguro, pp. 139–144; Rescher; and Russell (1900), Ch. IX.
13 Leibniz (1981); and Locke. 'Commentary' is to be understood in a somewhat tenuous sense. It was more a platform for Leibniz to present his own ideas.
14 Hobbes (1960), I, III; and Descartes (1984), II, p. 131.
15 *Philosophia Prima*, VII, 12 in Hobbes (1839–45), trans. José Benardete.
16 See the quotation at the beginning of Ch. 8 from Locke, II, 17, 20.
17 Locke, II, 17 and 29, 15–6. Note that these references correspond directly to those from Leibniz' commentary in note 12. See also note 13.
18 Hume (1888), I, II, I.
19 General references for the above are Hume (1888), I, II; Hume (1975), p. 124; and Berkeley (1962), I, 122–32.
20 Berkeley (1962), I, 123–8.
21 Hume (1888), I, II, IV.
22 See further Fogelin. See also below, Ch. 14, §3.

The Infinite

6 Kant

1 Much of the material in this chapter is derived from Moore (1988a), which contains more precise references.
2 Helpful background reading for this section are Allison; Strawson; and Warnock.
3 Kant (1964), pp. 121–2.
4 General references for this section are Kant (1933); Kant (1950); Kant (1964), Ch. III; and Kant (1956).
5 Kant (1933), B72.
6 Kant (1933), A19/B33. Cf. Kant (1933), B138–9.
7 This is a main theme of 'The Deduction of the Pure Concepts of Understanding' in Kant (1933), by common consent one of the most profound but also one of the most difficult passages in the whole of philosophy.
8 Kant (1933), A432/B460.
9 A general reference for the above is Kant (1933), *passim*. Heidegger especially emphasizes the importance of human finitude in Kant. See Heidegger (1962), *passim.*, esp. pp. 31 and 47; and Heidegger (1982), §14c.
10 See Kant (1933), A312–20/B368–77; and Kant (1956), I, II, II, VII.
11 The main sources for this section are 'Transcendental Dialectic' up to II, II and omitting II, I in Kant (1933); and Kant (1950), §§50–4.
12 Kant (1933), A32/B47–8. For discussion of various problems that arise here see Al-Azm, p. 18; Kemp Smith, pp. 95–8; Moore (1988a); and 'Infinity and Kant's Conception of the "Possibility of Experience"' in Parsons (1983). Note that space and time were *metaphysically* finite – mere conditions of one limited view of things. See Kant (1933), A27/B43 and A508/B536. Cf. how Nicholas of Cusa had regarded the 'privatively infinite' natural world as finite from God's point of view (see above, Ch. 3, §4).
13 Kant (1933), A503–5/B531–3.
14 Kant (1933), A670–2/B698–700.
15 Kant (1933), A486–8/B514–16.
16 Kant (1933), A426–43/B454–71.
17 This point was later taken up by Hegel. See Hegel (1969), pp. 243–4.
18 Kant (1933), B111 and A142–3/B182.
19 Kant (1933), A410/B437.
20 See further Bennett (1971); and Bennett (1974), Ch. 7. See also below, Ch. 14, §6.
21 See also below, note 22; and 'Zeno of Elea' in Bayle.
22 The P.S. of Leibniz' fourth paper and the N.B. of Clarke's fourth reply in Leibniz and Clarke. That Kant was principally concerned with these debates is convincingly argued in Al-Azm. See above, Ch. 5, note 9, for an important qualification concerning Newton himself.
23 Kant (1933), A466–70/B494–8. See also Benardete, p. 125. Kant's use of the term 'empiricist' here may force us to reassess his use of it in describing Aristotle (see above, Ch. 2, note 2).
24 Kant (1933), Bxx and A506–7/B534–5. See also Kant (1950), §50; Kant (1956), p. 111; and the letter to Garve in Kant (1902–55), XII, p. 258.
25 Cf. Strawson (1), pp. 203–6. See also Hawking, *passim*.
26 See further Al-Azm; Allison, esp. §3; Bennett (1971); Bennett (1974), Ch. 7; Broad, Ch.5, 1; Huby; Kemp Smith, on the 'Transcendental Dialectic'; 'Infinity and Kant's Conception of the "Possibility of Experience"' in Parsons (1983); and Strawson, Three, Chs. I and III.
27 Kant (1956), p. 166, translation adapted by me with the help of Geoffrey Plow and Graham White.

28 Kant (1933), Bxxx, his emphasis. See also Kant (1933), A641–2/B669–70 and B856ff.
29 Kant (1952), §§23–6.
30 Kant (1964), p. 104.
31 Cf. Kant (1933), A553–4/B581–2.
32 Kant (1960), pp. 32 and 60–1. See also Kant (1933), A828–9/B856–7 and Kant (1956), pp. 126–8.

7 Post-Kantian Metaphysics of the Infinite

1 Hegel (1969), pp. 228–34.
2 Hegel (1975), §62.
3 See Inwood, pp. 193ff.
4 Hegel (1975), §48.
5 Hegel (1969), pp. 234–8; and Hegel (1975), §§28 and 48.
6 Hegel (1975), §104.
7 Aristotle (1983), III, 6, 206a27–9 and 207a1.
8 Hegel (1969), p. 149.
9 Hegel (1975), §94.
10 General references for this section are Hegel (1969), pp. 116–57 and 225–38; and Hegel (1975), §§48, 92–5 and 104. See further Inwood, *passim*, esp. Chs. VI and VII; and C. Taylor, pp. 114–15 and 240–4.
11 Hamilton.
12 See Harries.
13 Von Hügel, *passim*. See also Kelly, pp. 165–7.
14 Bradley, esp. pp. 290–2. See also Royce, esp. §I.
15 'The One, the Many and the Infinite' in Royce.
16 A. E. Taylor, II, IV, 10 and III, IV, 6–9.
17 A. E. Taylor, II, III, 7.
18 See Dunkel.
19 Husserl, §143.
20 Bergson (1920), Ch. IV; and 'The Perception of Change' in Bergson (1975).
21 Quoted in Wang, p. 86.
22 McTaggart, Ch. XXII.
23 Wittgenstein (1961), 2.02ff. Wittgenstein's developed thoughts on infinity will be discussed in Ch. 9, §3. The *Tractatus* will come to the fore again in Ch. 13.
24 Kierkegaard, I, III, A. See also MacIntyre.
25 Jaspers (1950), Ch. III.
26 Third Lecture in Jaspers (1956). See further Koestenbaum.
27 Cf. Heidegger (1978), introduction, I.
28 'The Anaximander Fragment' in Heidegger (1975).
29 Heidegger (1962), *passim*, esp. pp. 31 and 47.
30 Heidegger (1982), §14c.
31 Cf. 'The Makropulos Case: Reflections on the Tedium of Immortality' in Williams (1973).
32 Heidegger (1978), pp. 329–31 and 424ff. and Heidegger (1982), p. 273. See further Krell, *passim*.
33 Sartre, p. 630.
34 Sartre, introduction.
35 Nietzsche (1961), II, 20.
36 See Čapek. See also Barnes (ed.) (1987), pp. 88 and 166–7.

37 Nietzsche (1967), §§1053–67.
38 'Towards New Seas' in Nietzsche (1974); Nietzsche (1974), §124; and Nietzsche (1961), III, 2. See further Nehamas, esp. Ch. 5.

8 The Mathematics of the Infinite, and the Impact of Cantor

1 A typical history of mathematics will contain surprisingly few references to the infinite. See e.g. Kline.
2 In §3 below this is precisely what we shall do.
3 The main source for this section is Bolzano.
4 Bolzano even extended this idea to the infinitely small, though, as I commented in Ch. 4, §2, the infinitely small was something he rejected in line with Cauchy.
5 'Proper' here indicates that we are disallowing the case of a subset that is the same as the original set.
6 'The Nature and Meaning of Numbers' in Dedekind.
7 He did not use this particular example however.
8 Useful background reading for §§2 and 3 is Bunn.
9 'The Nature and Meaning of Numbers' in Dedekind; and Peano.
10 General references for the above are Frege (1964); Frege (1967a); and Frege (1980).
11 Frege (1980), §§63 and 73.
12 See Hallett, p. 122.
13 Hume (1888), I, III, I.
14 Frege (1980), §§82–3.
15 I say '*the* empty set' because, given that a set is determined by its members, there cannot be more than one.
16 Russell (1967). Frege's reply is Frege (1967b).
17 Here I echo Dummett (1973), p. xxv.
18 Whitehead and Russell.
19 Russell (1919), pp. 138–9.
20 Whitehead and Russell, I, III, C.
21 Dummett has argued in Dummett (1967) that, quite apart from the paradoxes, Frege's own programme broke down at the point where he argued for the existence of infinitely many things. I confess I cannot see this.
22 Russell (1926), Ch. VII. General references for the above are Russell (1937), *passim*; Russell (1926), Chs. V–VII; and Russell (1958).
23 Quoted in Hallett, p. 25. Cf. the quotation in Rucker, p. 3.
24 See Tiles, p. 29 and cf. p. 69.
25 See Dauben (1980), pp. 216–19.
26 The main sources for this section are Cantor (1976) and Cantor (1955). Helpful background reading are Dauben (1979); Dauben (1980); Frænkel; Hallett; and Tiles.
27 Quoted in Dauben (1979), p. 55.
28 See Blanché.
29 See Dauben (1980), pp. 188–9.
30 For 'integers' here we can read 'natural numbers'.
31 Quoted in Dauben (1979), p. 1.
32 See Webb, pp. 129 and 131.
33 Peirce (1976). See further Hookway, pp. 176–80; and Murphey. Peirce also championed infinitesimals; see Peirce (1891–2).
34 Cantor (1976).
35 Cantor (1955).

Notes

36 Burali-Forti.
37 'ω' is the last letter of the Greek alphabet. It is pronounced 'omega'.
38 'ε' is the fifth letter of the Greek alphabet. It is pronounced 'epsilon'.
39 'Ω' is the capital version of 'ω'. See above, note 37.
40 Cantor (1967), p. 114.
41 Quoted in Hallett, p. 13.
42 Cf. the quotation in Rucker, p. 9. For discussion of the issues that are beginning
   to arise here see Hart (1975–6); Lear (1977); Maddy; Mayberry, p. 431; Moore
   (1985); 'What is the Iterative Conception of Set?' in Parsons (1983); and
   Rucker, esp. Ch. Five. See also below, Part Two, *passim*.
43 See Boolos (1971) and 'What is the Iterative Conception of Set?' in Parsons
   (1983). See also below, Ch. 10.
44 I am treating as single principles or axioms those schemata that would more
   usually be thought of as representing infinitely many axioms, all of a certain
   pattern.
45 Zermelo.
46 I am harmlessly overstating the generality of these principles. In fact they treat
   only of sets of a certain kind (see below, Ch. 10, §1).
47 Skolem (1967a); and Skolem (1967b).
48 That it can even be extended to Euclidean geometry is not trivial. Euclid's own
   work had to be refined and sharpened to show that it can. See Tarski.
49 Gödel (1967).

9 Reactions

1 The main sources for this section are Brouwer (1983a); and Brouwer (1983b).
2 Brouwer (1983a), p. 80.
3 See further Benacerraf and Putnam (ed.), introduction; Dummett (1977); 'The
  Philosophical Basis of Intuitionistic Logic' in Dummett (1978); Heyting
  (1983a); Heyting (1983b); and Percival (1986). See also below, Ch. 14, §§5 and 6.
4 The main source for this section is Hilbert.
5 See the quotation at the beginning of this chapter from Hilbert, p. 376.
6 On the possibility of finitely divisible space see Lucas (1984), pp. 76–8. On the
  possibility of finitely divisible time see Newton-Smith, Ch. VI.
7 Einstein, Ch. XXXI. For the most up-to-date views on the finitude of time and
  space see Hawking, *passim*.
8 Hilbert, p. 392.
9 This is the source that I have been using. See above, note 4.
10 Weyl (1967); and Weyl (1949), Ch. II.
11 Kaufmann.
12 Wittgenstein (1974a), II, 40.
13 This phrase was used by Wittgenstein in Wittgenstein (1975), p. 163.
14 Wittgenstein (1975), p. 159.
15 The 'sun' example is taken from Wittgenstein (1974b), §§350–1.
16 Wittgenstein (1975), p. 162.
17 This point is emphasized in Shanker, Ch. 5, 1, with references.
18 Wittgenstein (1974b), §124.
19 Wittgenstein (1978), V, 7. Cf. the two quotations at the beginning of this
   chapter from Wittgenstein (1976), p. 103, and Wittgenstein (1978), V, 7.
20 Gödel (1983).
21 Wittgenstein (1978), II, 19. This was not *exactly* Wittgenstein's example, but we
   can be confident that he would have said the same thing about it.
22 Wittgenstein (1974a), II, 40–1.

23 Wittgenstein (1976), p. 16.
24 Wittgenstein (1974a), II, 10.
25 Wittgenstein (1974a), II, 35 and 39.
26 The main sources for this section are Wittgenstein (1975), XII and pp. 304–14;
   Wittgenstein (1974a), II, 3, 7–10, 21, 35 and 39–43; Wittgenstein (1969), pp. 14–
   15 and 91–5; Wittgenstein (1976), pp. 16, 33, 103, 141, 255 and 269; and
   Wittgenstein (1978), II and V, *passim*. See further 'Wittgenstein's Philosophy of
   Mathematics' in Dummett (1978); Shanker, Ch. 5, 1; Wright (1979); and
   Wrigley, Ch. V.
27 'The Philosophical Basis of Intuitionistic Logic' in Dummett (1978).
28 The exegesis here is difficult. See 'Wittgenstein's Philosophy of Mathematics' in
   Dummett (1978); Kielkopf; and Wright (1979), esp. Ch. VII.
29 See e.g. van Dantzig; George; and Wright (1982).
30 'Wang's Paradox' in Dummett (1978).
31 Russell (1935–6), pp. 143–4. Cf. Weyl (1949), p. 42.
32 Benardete; and Rucker.
33 Quine (1969b), p. 62.

10 Transfinite Mathematics

1 Helpful background reading for this chapter are Enderton (1977); Crossley *et
   al.*, Ch. 6; Lieber; Maor; Quine (1969a); Rucker; J. Thomson (1967); and Tiles.
   Helpful background reading for Chapters 10–12 is Parsons (1967).
2 E.g. Benardete, p. 263; and Weyl (1949), p. 66.
3 Of course, we could exploit this notation in referring to the empty set itself, by
   labelling it '{}'. Thus instead of '{Ø}' we could write '{{}}'. This would
   register very graphically that we were dealing with sets whose existence
   depended on nothing but other sets. In more complex cases, however, such a
   notation would be unmanageably hard to read.
4 These are:
   Ø;
   {Ø};
   {{Ø}};
   {Ø, {Ø}};
   {{{Ø}}};
   {{Ø, {Ø}}};
   {Ø, {{Ø}}};
   {Ø, {Ø, {Ø}}};
   {{Ø}, {{Ø}}};
   {{Ø}, {Ø, {Ø}}};
   {{{Ø}}, {Ø, {Ø}}};
   {Ø, {Ø}, {{Ø}}};
   {Ø, {Ø}, {Ø, {Ø}}};
   {Ø, {{Ø}}, {Ø, {Ø}}};
   {{Ø}, {{Ø}}, {Ø, {Ø}}};
   and {Ø, {Ø}, {{Ø}}, {Ø, {Ø}}}.
5 Cf. Hart (1975–6) and 'What is the Iterative Conception of Set?' in Parsons
   (1983).
6 Lear (1977).
7 The concept of a Set and the concept of an ordinal are what Dummett has called
   indefinitely extensible concepts, and it is this feature of indefinite extensibility

which lies at the roots of these paradoxes. See 'The Philosophical Significance of Gödel's Theorem' in Dummett (1978), pp. 195–7.

8 See further Hart (1975–6); Lear (1977), pp. 86–7; Maddy; and Mayberry.

9 Von Neumann.

10 For discussion of the issues that arise here see Benacerraf (1965); and Resnik (1981).

11 '$\aleph$' is the first letter of the Hebrew alphabet. It is pronounced 'alef'.

12 The first of these is the first cardinal to succeed $\aleph_0, \aleph_{\aleph_0}, \aleph_{\aleph_\aleph}, \ldots$

13 There is a harmless ambiguity here. These uses of '+', '×' and so forth are different from their uses in the ordinal notation, eg in '$\omega + 1$'.

14 Hilbert maintained precisely that.

15 See Cohen.

16 Gödel (1983). Cf. above, Ch. 9, §3.

17 See Drake (1974).

18 Cohen, p. 151.

19 See further Smullyan.

20 See Rucker, pp. 81–3. A general reference for the above is Körner (1967).

21 I am not (here) denying, incidentally, that we can take all the points on a given line other than one of its end points, and collect them together into a set. What I am denying is that such a set would itself correspond to a line.

22 Rucker, p. 197.

11 The Löwenheim-Skolem Theorem

1 Much of the material in this chapter is derived from Moore (1985).

2 The main sources are Skolem (1967a); and Skolem (1967b). See further Boolos and Jeffrey, Ch. 13; Enderton (1972), §2.6; and Quine (1974), Ch. 32.

3 Skolem (1967b).

4 Although the solution is rigorous and uncontroversial, Skolem's paradox hooks up with various more general philosophical implications that the Löwenheim-Skolem theorem has, and there is a vast literature on it. See e.g. the exchange between Benacerraf and Wright in Benacerraf (1985) and Wright (1985); Fine; Hart (1970); Klenk; McIntosh; Putnam; Quine (1964); and the exchange between Resnik and Thomas in Resnik (1966), Thomas (1968), Resnik (1969) and Thomas (1971).

5 Skolem (1967b), pp. 295–6.

6 Putnam, p. 444. My position in this section is very close to Putnam's.

7 This is an allusion to the famous last sentence, numbered 7, of Wittgenstein (1961). See below, Ch. 13, for further bearing of the *Tractatus* on this.

12 Gödel's Theorem

1 But see above, Ch. 8, note 48. This possibility was established only recently, using techniques that go beyond what was in Euclid's own work.

2 Cf. Frege (1980), §5. For one of the best known attempts to apply the paradigm to arithmetic see Peano.

3 See above, Ch. 8, note 44. I count schemata as single axioms.

4 The main source for this section is Gödel (1967). There is a crucial embellishment to Gödel's result in Rosser. For presentations of the result, of varying degrees of detail and accessibility, see Boolos and Jeffrey, Chs 14–16; Crossley *et al.*, Ch. 5; Enderton (1972), Ch. Three; van Heijenoort (1967a); Hofstadter, *passim*; Lucas (1970), §24; Quine (1974), Ch. 33; and

Rucker, Exc. Two. Most helpful probably, combining a high degree of detail with a high degree of accessibility, is Nagel and Newman.

5 Nagel and Newman, pp. 94–5.
6 Compare this with the quotation from Gödel with which I opened this chapter. This is quoted in Wang, p. 324.
7 Gödel (1967), p. 615.
8 Detlefsen, in Detlefsen (1986), has provided a sustained and fascinating critique of what he calls Hilbert's programme, which shows that it is *not* threatened by Gödel's theorem, and that it is still very much a live option. I cannot possibly do justice to his arguments here. Suffice to say that what he understands by Hilbert's programme is not what I understand by it; nor, I think, is it anywhere near as ambitious as what Hilbert himself had in mind. Certainly, anyone content with what Detlefsen presents as ideal methods of proof (whose *raison d'être* is a matter of efficiency and utility) would long since have vacated 'Cantor's paradise'.
9 Lucas (1961); and Lucas (1970), §§16–30. He even takes the theorem to refute a kind of physical determinism.
10 See e.g. Benacerraf (1967); 'The Abilities of Men and Machines' in Dennett; Hofstadter, esp. pp. 471–7 and 577–8; and Webb, *passim*.
11 This quotation, emphasis Gödel's, comes shortly after that with which I opened this chapter. See above, note 6. For a fuller discussion of these issues see the pieces cited in note 10, and for Lucas' rejoinders to some of the points that have been made against him see Lucas (1970), §26.
12 I have tried to defend these ideas in greater depth in Moore (1988b).
13 'The Philosophical Significance of Gödel's Theorem' in Dummett (1978), p. 198.
14 Wittgenstein (1974b), §195.
15 Cf. Wittgenstein (1974b), §208. See also Wittgenstein (1974b), §229.

13 Saying and Showing

1 See further below, Ch. 14, §§2 and 4 and Ch. 15, §3.
2 Wittgenstein (1961).
3 1.
4 2.0122, his emphasis.
5 6.54 and 7. Cf the end of Ch. 9 above.
6 Some of this comes through more clearly in his preparatory notebooks, Wittgenstein (1979), than in the *Tractatus*.
7 Engelmann, pp. 143–4, Wittgenstein's emphasis.
8 5.633–5.6331. I am going some way beyond what is actually in the text.
9 For the influence of Schopenhauer on Wittgenstein's thinking in all of this see below, Ch. 15, §3.
10 For a more thorough treatment of the issues in this section see Moore (1987).
11 I am harmlessly ignoring certain complications that arise because of words such as 'I', 'here', and 'now', whose meaning is context-sensitive.
12 p. xx.
13 6.522 and 6.45.
14 6.4311. A general reference for the above is 6.43ff.
15 6.4312.
16 We are reminded again of the Pythagorean and Platonic idea of the imposition of the *peras* on to *apeiron* (cf above, Ch. 7, §4). For more on these themes see below, Ch. 15, §4. Cf. also Plotinus I, 5, 7.
17 Wittgenstein (1975), p. 157, his emphasis.

18 Wittgenstein (1975), p. 164.
19 Wrigley, esp. Ch. V.
20 Wittgenstein (1974a), p. 282, his emphasis.
21 Wittgenstein (1974b), §229.
22 Cf. 'Wittgenstein and Idealism' in Williams (1981); and Moore (1987), §8. (Although he was not continuing to use the word 'show' in exactly its *Tractatus* sense, it is interesting to observe what he went on to say about the expression 'and so on', immediately after having said that it was nothing but the expression 'and so on': '[It is] nothing, that is, but a sign in a calculus which can't do more than have meaning *via* the rules that hold of it; which can't say more than it shows.' (Wittgenstein (1974a), p. 282. Cf. p. 469.))
23 Cf. Mayberry, p. 431.
24 Rucker, Ch. Five, contains ideas very close to these. Cf. esp. p. 205, where he refers to the *Tractatus*. Cf. also p. 191, where he says, 'Rationally the universe is a Many, but mystically it is a One.'

14 Infinity Assessed. The History Reassessed

1 *Parmenides*, 132–4 in Plato.
2 Cf. H. Owen.
3 Wittgenstein (1961), 6.431. Cf. Heidegger (1978), p. 237.
4 Kant (1933), A508–10/B536–8.
5 See above, Ch. 6, note 12. If the framework is construed in a Heideggerian way, it is mathematically finite.
6 Hume (1888).
7 Strawson, p. 162.
8 Compare the ideas in this section with Hart (1975–6); and 'What is the Iterative Conception of Set?' in Parsons (1983).
9 I am here repeating an earlier characterization. See above, p. 81.
10 Kant (1933), A498/B526, his emphasis.
11 Cf. Lear (1988), p. 67. §§3–4 are generally helpful here. But I think he underestimates the force of the crude reading.
12 See eg. *Physics*, III, 7, 207b10–15 in Aristotle (1984).
13 *Physics*, VI, 11, 219b6 in Aristotle (1984).
14 Wittgenstein (1975), pp. 160–1.
15 E.g. Wittgenstein (1974b), §194.
16 Compare what follows with Lear (1977).
17 But see below, §6, for qualifications concerning Brouwer.
18 Kant (1933), A503–5/B531–3.
19 I think this explains why Dummett, when exploring similar issues in 'The Reality of the Past' in Dummett (1978) spells out a motivation for what he calls 'anti-realism' in terms which the anti-realist ought to reject. See pp. 369–70.
20 Most of the material for this section is derived from Moore (1989–90).
21 Brouwer (1983a).
22 See Clark and Read.
23 The term 'super-task' was first introduced in J. Thomson (1954).
24 Cf. Wright (1982), p. 248.
25 Wright (1982), p. 248.
26 Cf. Wittgenstein (1975), pp. 307–8.
27 Cf. Wittgenstein (1975), p. 159.
28 For an interesting echo of the ideas in this paragraph see Geach, appendix. I am committed to regarding Geach's suggestion as incoherent.
29 See Lear (1979–80); and Lear (1988), pp. 82–3.

30 Cf. Bennett (1971), p. 135; and Bennett (1974), p. 122.
31 For discussion of how the pressure is felt in modern physics, and for ways of dealing with it, see Hawking, *passim*, esp. Ch. 10.

15 Human Finitude

1 Compare the ideas in this chapter with much of Barrett; and with Levinas.
2 Cf. T. Nagel, Ch. III.
3 I. Murdoch, p. 36. For a marvellous religious expression of a similar experience, where human finitude is at the same time set against the infinitude of God, see Psalm CXXXIX.
4 Cf. Wittgenstein (1979), p. 74.
5 Hawking, p. 136.
6 Hawking, p. 174.
7 Kant (1933), Aviii. Cf. Kant (1933), A700–2/B728–30.
8 See above, Ch. 6, note 7.
9 Kant (1933), B130–1.
10 Wittgenstein (1961), 5.631.
11 'The Paralogisms of Pure Reason' in Kant (1933).
12 For an interesting insight into how they are related see Kant (1933), A398.
13 Wittgenstein (1961), 5.62–5.63.
14 Cf. in this connection Kant (1933), A25/B39 and A32/B47–8.
15 Cf. Moore (1987), §6.
16 Kant (1933), A129.
17 See Hacker, pp. 87–100; and Janik and Toulmin, *passim*.
18 See Hacker, p. 88.
19 Schopenhauer (1969), IV, §54, his emphasis. Cf. 'On the Indestructibility of Our Essential Being by Death' in Schopenhauer (1970). Here it is instructive to think once again of the Parmenidean sphere.
20 Wittgenstein (1961), 6.4311.
21 Compare the ideas throughout this section with those of the existentialists (see above, Ch. 7, §4).
22 It is not obvious why there should be this asymmetry between my future non-existence and my past non-existence. See T. Nagel, pp. 328–9.
23 Wittgenstein (1961), 6.4312. Wittgenstein's question was rhetorical, but see above, Ch. 6, §4 for a Kantian answer.
24 Cf. 'The Makropulos Case: Reflections on the Tedium of Immortality' in Williams (1973).
25 T. Nagel, p. 224.
26 Wittgenstein (1961), 6.4311.
27 See G. Thomson, Ch. III.
28 For a superb discussion of death and its harmfulness, see T. Nagel, Ch. XI, §3.
29 It is instructive here to compare the quotation from Schnell in Amis, p. 16.
30 This echoes part of the quotation which I have used as a frontispiece for this book, taken from Kant (1956), p. 166.
31 Cf. St Luke, XVII, 33 and Romans, VI–VIII, *passim*. It is interesting also to reflect on how this bears on Wittgenstein (1961), 5.64: 'Here it can be seen that solipsism, when its implications are followed out strictly, coincides with pure realism. The self of solipsism shrinks to a point without extension, and there remains the reality co-ordinated with it.'
32 Cf. Wittgenstein (1979), p. 75. See also above, Ch. 13, §3. Cf. also some of Hegel; see Inwood, Ch. XI, §4.

33 For links with Wittgenstein, see above, Ch. XIII, §3.
34 I Corinthians, XIII, 13.
35 Kant (1933), A804–5/B832–3.
36 Wittgenstein (1979), p. 74.

Bibliography

Al-Azm, Sadik J. (1972) *The Origins of Kant's Arguments in The Antinomies*, Oxford University Press.

Allison, Henry E. (1983) *Kant's Transcendental Idealism*, Yale University Press.

Amis, Martin (1988) *Einstein's Monsters*, Harmondsworth, Penguin.

Aquinas, St Thomas (1947) *Summa Theologica*, trans. the Fathers of the English Dominican Province, New York.

Aristotle (1983) *Physics, Books III and IV*, trans. Edward Hussey, Oxford University Press.

Aristotle (1984) *The Complete Works*, ed. Jonathan Barnes, Princeton University Press.

Augustine, St (1945) *The City of God*, trans. John Healey, London, Dent.

Augustine, St (1961) *Confessions*, trans. R. F. Pine-Coffin, Harmondsworth, Penguin.

Barnes, Jonathan (1979) *The Presocratic Philosophers*, London, Routledge & Kegan Paul.

Barnes, Jonathan (ed.) (1987) *Early Greek Philosophy*, Harmondsworth, Penguin.

Barrett, William (1987) *Death of the Soul: From Descartes to the Computer*, Oxford University Press.

Bayle, Pierre (1965) *Historical and Critical Dictionary*, ed. and trans. Richard H. Popkin, Indianapolis, Bobbs-Merrill.

Benacerraf, Paul (1962) 'Tasks, Supertasks and the Modern Eleatics', in *Journal of Philosophy*, 59.

Benacerraf, Paul (1965) 'What Numbers Could Not Be', in *Philosophical Review*, 74.

Benacerraf, Paul (1967) 'God, The Devil and Gödel', in *The Monist*, 51.

Benacerraf, Paul (1985) 'Skolem and the Skeptic', in *Proceedings of the Aristotelian Society*, Supp. Vol. 59.

Benacerraf, Paul and Hilary Putnam (eds) (1983) *Philosophy of Mathematics: Selected Readings*, Cambridge University Press.

Benardete, José A. (1964) *Infinity: An Essay in Metaphysics*, Oxford University Press.

Bennett, Jonathan (1971) 'The Age and Size of the World', in *Synthèse*, 23.

Bennett, Jonathan (1974) *Kant's Dialectic*, Cambridge University Press.

Bennett, Jonathan (1984) *A Study of Spinoza's 'Ethics'*, Cambridge University Press.

Beresford, Geoffrey C. (1980) 'A Note on Thomson's Lamp "Paradox"', in *Analysis*, 40.

Bergson, Henri (1920) *Creative Evolution*, trans. Arthur Mitchell, London, Macmillan.

250

Bibliography

Bergson, Henri (1975) *An Introduction to Metaphysics: The Creative Mind*, trans. Mabelle L. Andison, Totowa, Littlefield, Adams & Co..

Berkeley, George (1901) *The Works*, ed. A. C. Fraser, Oxford University Press.

Berkeley, George (1962) *The Principles of Human Knowledge*, ed. G. J. Warnock, London, Fontana.

Black, Max (1950–1) 'Achilles and the Tortoise', in *Analysis,* 11.

Blanché, Robert (1967) 'Couturat', trans. Albert E. Blumberg, in Paul Edwards (ed.) *The Encyclopædia of Philosophy,* London, Macmillan.

Bolzano, Bernard (1950) *Paradoxes of the Infinite*, trans. F. Prihonsky, London, Routledge & Kegan Paul.

Boolos, George S. (1971) 'The Iterative Conception of a Set', in *Journal of Philosophy,* 68.

Boolos, George S. and Richard C. Jeffrey (1980) *Computability and Logic*, Cambridge University Press.

Bos, H. J. M. (1980) 'Newton, Leibniz and the Leibnizian Tradition', in I. Grattan-Guiness (ed.), *From the Calculus to Set Theory, 1630–1910,* London, Duckworth.

Bostock, David (1972–3) 'Aristotle, Zeno and the Potential Infinite', in *Proceedings of the Aristotelian Society,* 73.

Boyer, Carl B. (1954) *The History of the Calculus and its Conceptual Development*, New York, Dover.

Bradley, F. H. (1897) *Appearance and Reality*, London.

Broad, C. D. (1978) *Kant: An Introduction*, ed. C. Lewy, Cambridge University Press.

Brouwer, L. E. J. (1983a) 'Intuitionism and Formalism', in Paul Benacerraf and Hilary Putnam (eds), *Philosophy of Mathematics,* Cambridge University Press.

Brouwer, L. E. J. (1983b) 'Consciousness, Philosophy and Mathematics', in Paul Benacerraf and Hilary Putnam (eds), *Philosophy of Mathematics,* Cambridge University Press.

Bruno, Giordano (1950) *On the Infinite Universe and Worlds*, in D. W. Singer *Giordano Bruno*, New York.

Bunn, R. (1980) 'Developments in the Foundations of Mathematics, 1870–1910', in I. Grattan-Guiness (ed.), *From the Calculus to Set Theory, 1630–1910,* London, Duckworth.

Burali-Forti, Cesare (1967) 'A Question on Transfinite Numbers', trans. Jean van Heijenoort, in Jean van Heijenoort (ed.), *From Frege to Gödel: A Source Book in Mathematical Logic, 1879–1931,* Harvard University Press.

Cantor, Georg (1932) *Gesammelte Abhandlungen*, ed. E. Zermelo, Berlin.

Cantor, Georg (1955) *Contributions to the Founding of the Theory of Transfinite Numbers*, trans. Philip E.B. Jourdain, New York, Dover.

Cantor, Georg (1967) 'Letter to Dedekind', trans. Jean van Heijenoort and Stefan Bauer-Mengelberg, in Jean van Heijenoort (ed.), *From Frege to Gödel: A Source Book in Mathematical Logic, 1879–1931,* Harvard University Press.

Cantor, Georg (1976) 'Foundations of the Theory of Manifolds', trans. U. Parpart, in *The Campaigner,* 9.

Čapek, Milič (1967) 'Eternal Return', in Paul Edwards (ed.), *The Encyclopædia of Philosophy,* London, Macmillan.

Chihara, C. (1965) 'On the Possibility of Completing an Infinite Process', in *Philosophical Review,* 74.

Clark, Peter and Stephen Read (1984) 'Hypertasks', in *Synthèse,* 61.

Cohen, P. J. (1966) 'Set Theory and the Continuum Hypothesis', New York, Benjamin.

Crossley, J. N. et al (1972) *What is Mathematical Logic?*, Oxford University Press.

van Dantzig, D. (1955) 'Is $10^{10^{10}}$ a Finite Number?', in *Dialectica,* 9.

Dauben, Joseph W. (1979) *Georg Cantor: His Mathematics and Philosophy of the Infinite*, Harvard University Press.

Dauben, Joseph W. (1980) 'The Development of Cantorian Set Theory', in I. Grattan-Guiness (ed.), *From the Calculus to Set Theory, 1630–1910*, London, Duckworth.

Dedekind, Richard (1901) *Essays on the Theory of Numbers*, trans. W. W. Beman, Chicago, Open Court.

Dennett, Daniel (1981) *Brainstorms: Philosophical Essays on Mind and Psychology*, Sussex, Harvester.

Descartes, René (1964–75) *Oeuvres*, ed. Charles Adam and Paul Tannery, Paris, Vrin.

Descartes, René (1984) *The Philosophical Writings*, trans. John Cottingham, Robert Stoothoff, and Dugald Murdoch, Cambridge University Press.

Detlefsen, Michael (1986) *Hilbert's Program: An Essay on Mathematical Instrumentalism*, Dordrecht, Reidel.

Diels, Hermann (ed.) (1882–1909) *Commentaria in Aristotelem Graeca*, Berlin.

Drake, F. R. (1974) *Set Theory: An Introduction to Large Cardinals*, Amsterdam, North Holland.

Dretske, Fred I. (1964-5) 'Counting to Infinity', in *Analysis*, 25.

Duhem, Pierre (1985a) *Mediæval Cosmology: Theories of Infinity, Place, Time, Void and the Plurality of Worlds*, trans. Roger Ariew, Chicago University Press.

Duhem, Pierre (1985b) *To Save the Phenomena: An Essay on the Idea of Physical Theory from Plato to Galileo*, trans. Edmund Dolan and Chaninah Maschler, Chicago University Press.

Dummett, Michael (1967) 'Frege', in Paul Edwards (ed.), *The Encyclopædia of Philosophy*, London, Macmillan.

Dummett, Michael (1973) *Frege: Philosophy of Language*, London, Duckworth.

Dummett, Michael (1977) *Elements of Intuitionism*, Oxford University Press.

Dummett, Michael (1978) *Truth and Other Enigmas*, London, Duckworth.

Dunkel, Harold B. (1967) 'Herbart', in Paul Edwards (ed.), *The Encyclopædia of Philosophy*, London, Macmillan.

Edwards, Paul (ed.) (1967) *The Encyclopædia of Philosophy*, London, Macmillan.

Einstein, Albert (1920) *Relativity: The Special and General Theory*, trans. Robert W. Lawson, London, Methuen.

Emmett, E. R. (1957) 'Infinity', in *Mind*, 66.

Enderton, Herbert B. (1972) *A Mathematical Introduction to Logic*, New York, Academic Press.

Enderton, Herbert B. (1977) *Elements of Set Theory*, New York, Academic Press.

Engelmann, Paul (1967) *Letters from Ludwig Wittgenstein, with a Memoir*, ed. B. F. McGuinness and trans. L. Furtmüller, Oxford, Blackwell.

Euclid (1956) *Elements*, ed. T.L. Heath, New York, Dover.

Euler, Leonhard (1748) *Introduction in Analysin Infinitorum*, Lausanne.

Findlay, J. N. (1953) 'The Notion of Infinity', in *Proceedings of the Aristotelian Society*, Supp. Vol. 27.

Fine, Arthur (1968) 'Quantification Over the Real Numbers', in *Philosophical Studies*, 19.

Fogelin, Robert (1988) 'Hume and Berkeley on the Proofs of Infinite Divisibility', in *Philosophical Review*, 97.

Frænkel, Abraham A. (1967) 'Cantor', in Paul Edwards (ed.), *The Encyclopædia of Philosophy*, London, Macmillan.

Frege, Gottlob (1964) *The Basic Laws of Arithmetic: Exposition of the System*, trans. Montgomery Furth, University of California Press.

Bibliography

Frege, Gottlob (1967a) 'Begriffsschrift, A Formula Language, Modelled Upon That of Arithmetic, for Pure Thought', trans. Stefan Bauer-Mengelberg, in Jean van Heijenoort (ed.) *From Frege to Gödel: A Source Book in Mathematical Logic, 1879–1931*, Harvard University Press.

Frege, Gottlob (1967b) 'Letter to Russell', trans. Beverly Woodward, in Jean van Heijenoort (ed.), *From Frege to Gödel: A Source Book in Mathematical Logic, 1879–1931*, Harvard University Press.

Frege, Gottlob (1980) *The Foundations of Arithmetic: A Logico-Mathematical Inquiry Into the Concept of Number*, trans. J. L. Austin, Oxford, Blackwell.

Furley, David J. (1967) 'Parmenides', in Paul Edwards (ed.), *The Encyclopædia of Philosophy*, London, Macmillan.

Gale, Richard M. (ed.) (1968) *The Philosophy of Time*, Sussex, Harvester.

Galileo Galilei (1914) *Dialogues Concerning Two New Sciences*, trans. Henry Crew and Alfonso de Salvio, New York, Dover.

Geach, Peter (1977) *Providence and Evil*, Cambridge University Press.

George, Alexander (1988) 'The Conveyability of Intuitionism, an Essay on Mathematical Cognition', in *Journal of Philosophical Logic,* 17.

Gödel, Kurt (1967) 'On Formally Undecidable Propositions of *Principia Mathematica* and Related Systems I', trans. Jean van Heijenoort, in Jean van Heijenoort (ed.), *From Frege to Gödel: A Source Book in Mathematical Logic, 1879–1931*, Harvard University Press.

Gödel, Kurt (1983) 'What is Cantor's Continuum Problem?', in Paul Benacerraf and Hilary Putnam (eds), *Philosophy of Mathematics,* Cambridge University Press.

Grattan-Guiness, I. (ed.) (1980a) *From the Calculus to Set Theory, 1630–1910*, London, Duckworth.

Grattan-Guiness, I. (1980b) 'The Emergence of Mathematical Analysis and its Foundational Progress, 1780–1880', in I. Grattan-Guiness (ed.) *From the Calculus to Set Theory, 1630–1910*, London, Duckworth.

Grünbaum, Adolf (1968a) 'Modern Science and Zeno's Paradoxes of Motion', in Richard M. Gale (ed.), *The Philosophy of Time*, Sussex, Harvester.

Grünbaum, Adolf (1968b) *Modern Science and Zeno's Paradoxes*, London, Allen & Unwin.

Grünbaum, Adolf (1973) *Philosophical Problems of Space and Time*, Dordrecht, Reidel.

Hacker, P. M. S. (1986) *Insight and Illusion: Themes in the Philosophy of Wittgenstein*, Oxford University Press.

Hallett, Michael (1984) *Cantorian Set Theory and Limitation of Size*, Oxford University Press.

Hamilton, William (1829) 'On the Philosophy of the Unconditioned', in *Edinburgh Review,* 1.

Harré, H. R. (1964) 'Infinity', in *Proceedings of the Aristotelian Society,* Supp. Vol. 38.

Harries, Karsten (1967) 'Solger', in Paul Edwards (ed.), *The Encyclopædia of Philosophy*, London, Macmillan.

Hart, W. D. (1970) 'Skolem's Promises and Paradoxes', in *Journal of Philosophy,* 67.

Hart, W. D. (1975–6) 'The Potential Infinite', in *Proceedings of the Aristotelian Society,* 76.

Hawking, Stephen W. (1988) *A Brief History of Time: From the Big Bang to Black Holes*, London, Bantam Press.

Heath, T. L. (1921) *A History of Greek Mathematics*, Oxford University Press.

The Infinite

Hegel, G. W. F. (1969) *Science of Logic*, trans. A.V. Miller, London, Allen & Unwin.

Hegel, G. W. F. (1975) *The Encyclopædia of the Philosophical Sciences*, Part One ('Logic') trans. William Wallace, Oxford University Press.

Heidegger, Martin (1962) *Kant and the Problem of Metaphysics*, trans. James S. Churchill, Indiana University Press.

Heidegger, Martin (1975) *Early Greek Thinking: The Dawn of Western Philosophy*, trans. David Farrell Krell and Frank A. Capuzzi, San Fransisco, Harper & Row.

Heidegger, Martin (1978) *Being and Time*, trans. John Macquarrie and Edward Robinson, Oxford, Blackwell.

Heidegger, Martin (1982) *The Basic Problems of Phenomenology*, trans. Albert Hofstadter, Indiana University Press.

van Heijenoort, Jean (1967) 'Gödel's Theorem', in Paul Edwards (ed.), *The Encyclopædia of Philosophy*, London, Macmillan.

van Heijenoort, Jean (ed.) (1967) *From Frege to Gödel: A Source Book in Mathematical Logic, 1879–1931*, Harvard University Press.

Henkin, L. P. Suppes, and A. Tarski (eds) (1959) *The Axiomatic Method, with Special Reference to Geometry and Physics*, Amsterdam, North Holland.

Heyting, Arend (1983a) 'The Intuitionist Foundations of Mathematics', trans. Erna Putnam and Gerald J. Massey, in Paul Benacerraf and Hilary Putnam (eds), *Philosophy of Mathematics*, Cambridge University Press.

Heyting, Arend (1983b) 'Disputation', in Paul Benacerraf and Hilary Putnam (eds), *Philosophy of Mathematics*, Cambridge University Press.

Hilbert, David (1967) 'On the Infinite', trans. Stefan Bauer-Mengelberg, in Jean van Heijenoort (ed.), *From Frege to Gödel: A Source Book in Mathematical Logic, 1879–1931*, Harvard University Press.

Hintikka, Jaakko (1957) 'Necessity, Universality and Time in Aristotle', in *Ajatus*, 20.

Hintikka, Jaakko (1966) 'Aristotelian Infinity', in *Philosophical Review*, 75.

Hobbes, Thomas (1839–45) *Opera Philosophica*, ed. William Molesworth, London.

Hobbes, Thomas (1960) *Leviathan, or the Matter, Form and Power of a Commonwealth Ecclesiastical and Civil*, ed. Michael Oakeshott, Oxford, Blackwell.

Hofstadter, Douglas R. (1979) *Gödel, Escher, Bach: An Eternal Golden Braid, A Metaphorical Fugue on Minds and Machines, in the Spirit of Lewis Carroll*, Harmondsworth, Penguin.

Hookway, Christopher (1985) *Peirce*, London, Routledge & Kegan Paul.

de l'Hôpital, G. F. A. (1715) *Analyse des Infiniment Petites Pour l'Intelligence des Lignes Courbes*, Paris.

Huby, Pamela M. (1971) 'Kant or Cantor?: That the Universe, if Real, Must be Finite in Both Space and Time', in *Philosophy*, 46.

von Hügel, Baron Friedrich (1908) *The Mystical Element of Religion as Studied in Saint Catherine of Genoa and Her Friends*, London.

Hume, David (1888) *A Treatise of Human Nature*, ed. L. A. Selby-Bigge, Oxford University Press.

Hume, David (1975) *Enquiries Concerning Human Understanding and Concerning the Principles of Morals*, ed. L. A. Selby-Bigge, Oxford University Press.

Husserl, Edmund (1931) *Ideas: General Introduction to Pure Phenomenology*, trans. W. R. Boyce Gibson, London, Allen & Unwin.

Inwood, M. J. (1983) *Hegel*, London, Routledge & Kegan Paul.

Ishiguro, Hidé (1972) *Leibniz's Philosophy of Logic and Language*, London, Duckworth.

Bibliography

Janik, Allan and Stephen Toulmin (1973) *Wittgenstein's Vienna*, New York, Touchstone.

Jaspers, Karl (1950) *The Perennial Scope of Philosophy*, trans. Ralph Manheim, London, Routledge & Kegan Paul.

Jaspers, Karl (1956) *Reason and Existenz*, trans. William Earle, London, Routledge & Kegan Paul.

Kant, Immanuel (1902–55) *Gesammelte Schriften*, Berlin.

Kant, Immanuel (1933) *Critique of Pure Reason*, trans. Norman Kemp Smith, London, Macmillan.

Kant, Immanuel (1950), *Prolegomena to Any Future Metaphysics*, trans. Lewis White Beck, Indianapolis, Bobbs-Merrill.

Kant, Immanuel (1952) *Critique of Judgement*, trans. James Creed Meredith, Oxford University Press.

Kant, Immanuel (1956) *Critique of Practical Reason*, trans. Lewis White Beck, Indianapolis, Bobbs-Merrill.

Kant, Immanuel (1960) *Religion Within the Limits of Reason Alone*, trans. Theodore M. Greene and Hoyt H. Hudson, New York, Harper & Row.

Kant, Immanuel (1964), *Groundwork of the Metaphysics of Morals*, trans. H.J. Paton, New York, Harper & Row.

Kaufmann, Felix (1978) *The Infinite in Mathematics: Logico-Mathematical Writings*, ed. B.F. McGuinness and trans. Paul Foulkes, Dordrecht, Reidel.

Kelly, James J. (1983) *Baron Friedrich von Hügel's Philosophy of Religion*, Leuven University Press.

Kemp Smith, Norman (1923) *A Commentary to Kant's 'Critique of Pure Reason'*, London, Macmillan.

Kerferd, G.B. (1967) 'Apeiron/Peras', in Paul Edwards (ed.), *The Encyclopædia of Philosophy*, London, Macmillan.

Kielkopf, Charles F. (1970) *Strict Finitism: An Examination of Ludwig Wittgenstein's 'Remarks on the Foundations of Mathematics'*, The Hague, Mouton.

Kierkegaard, Søren (1954) *The Sickness Unto Death*, trans. Walter Lowrie, Princeton University Press.

Klenk, Virginia (1976) 'Intended Models and the Löwenheim-Skolem Theorem', in *Journal of Philosophical Logic*, 5.

Kline, Morris (1972), *Mathematical Thought From Ancient to Modern Times*, Oxford University Press.

Kneale, William C. (1967) 'Eternity', in Paul Edwards (ed.), *The Encyclopædia of Philosophy*, London, Macmillan.

Knorr, Wilbur R. (1982) 'Infinity and Continuity: The Interaction of Mathematics and Philosophy in Antiquity', in Norman Kretzmann (ed.), *Infinity and Continuity in Ancient and Mediæval Thought*, Cornell University Press.

Koestenbaum, Peter (1967) 'Jaspers', in Paul Edwards (ed.), *The Encyclopædia of Philosophy*, London, Macmillan.

Körner, S. (1953) 'The Notion of Infinity', in *Proceedings of the Aristotelian Society*, Supp. Vol. 27.

Körner, S. (1967) 'Continuity', in Paul Edwards (ed.), *The Encyclopædia of Philosophy*, London, Macmillan.

Koyré, Alexandre (1957) *From the Closed World to the Infinite Universe*, Baltimore, The Johns Hopkins Press.

Krell, David Farrell (1986) *Intimations of Mortality: Time, Truth and Finitude in Heidegger's Thinking of Being*, The Pennsylvania State University Press.

Kretzmann, Norman (ed.) (1982) *Infinity and Continuity in Ancient and Mediæval Thought*, Cornell University Press.

The Infinite

Kretzmann, Norman, Anthony Kenny, and Jan Pinborg (eds) (1982) *The Cambridge History of Later Mediæval Philosophy*, Cambridge University Press.
Lakatos, Imre (ed.) (1967) *Problems in the Philosophy of Mathematics*, Amsterdam, North Holland.
Lang, Serge (1968) *A First Course in Calculus*, London, Addison-Wesley.
Lasserre, François (1964) *The Birth of Mathematics in the Age of Plato*, London, Hutchinson.
Lear, Jonathan (1977) 'Sets and Semantics', in *Journal of Philosophy*, 74.
Lear, Jonathan (1979–80) 'Aristotelian Infinity', in *Proceedings of the Aristotelian Society*, 80.
Lear, Jonathan (1981), 'A Note on Zeno's Arrow', in *Phronesis*, 27.
Lear, Jonathan (1982) 'Aristotle's Philosophy of Mathematics', in *Philosophical Review*, 91.
Lear, Jonathan (1988) *Aristotle: The Desire to Understand*, Cambridge University Press.
Leibniz, G.W. (1960–1) *Philosophische Schriften*, Hildesheim.
Leibniz, G.W. (1962) *Mathematische Schriften*, Hildesheim.
Leibniz, G.W. (1981) *New Essays on Human Understanding*, trans. Peter Remnant and Jonathan Bennett, Cambridge University Press.
Leibniz, G.W. and Samuel Clarke (1956), *Correspondence*, ed. H.G. Alexander, Manchester University Press.
Levinas, Emmanuel (1985) *Ethics and Infinity: Conversations with Philippe Nemo*, trans. Richard A. Cohen, Duquesne University Press.
Lewy, C. (1953) 'The Notion of Infinity', in *Proceedings of the Aristotelian Society*, Supp. Vol. 27.
Lieber, Lillian R. (1953) *Infinity*, New York, Holt, Rinehart & Winston.
Locke, John (1965) *Essay Concerning Human Understanding*, ed. John Yolton, London, Dent.
Lovejoy, Arthur O. (1936) *The Great Chain of Being*, Harvard University Press.
Lucas, J.R. (1961) 'Minds, Machines and Gödel', in *Philosophy*, 36.
Lucas, J.R. (1970) *The Freedom of the Will*, Oxford University Press.
Lucas, J.R. (1984) *Space, Time and Causality: An Essay in Natural Philosophy*, Oxford University Press.
McIntosh, Clifton (1979), 'Skolem's Criticisms of Set Theory', in *Noûs*, 13.
MacIntyre, Alasdair (1967) 'Kierkegaard', in Paul Edwards (ed.), *The Encyclopædia of Philosophy*, London, Macmillan.
McTaggart, J.M.E. (1921) *The Nature of Existence*, Cambridge University Press.
Maddy, Penelope (1983) 'Proper Classes', in *Journal of Symbolic Logic*, 48.
Magee, Bryan (ed.) (1987) *The Great Philosophers*, London, BBC Books.
Maor, Eli (1986) *To Infinity and Beyond: A Cultural History of the Infinite*, Stuttgart, Birkhäuser.
Mayberry, John (1986) 'Review of Michael Hallett's *Cantorian Set Theory and Limitation of Size*', in *The Philosophical Quarterly*, 36.
Moore, A.W. (1985) 'Set Theory, Skolem's Paradox and the *Tractatus*', in *Analysis*, 45.
Moore, A.W. (1987) 'On Saying and Showing', in *Philosophy*, 62.
Moore, A.W. (1988a) 'Aspects of the Infinite in Kant', in *Mind*, 97.
Moore, A.W. (1988b) 'What Does Gödel's Second Incompleteness Theorem Show?', in *Noûs*, 22.
Moore, A.W. (1989) 'Early Greek Philosophers on the Infinite', in *Cogito*, 3.
Moore, A.W. (1989–90) 'A Problem for Intuitionism: The Apparent Possibility of Performing Infinitely Many Tasks in a Finite Time', in *Proceedings of the Aristotelian Society*, 90.

Bibliography

Murdoch, Iris (1977) *The Fire and the Sun: Why Plato Banished the Artist*, Oxford University Press.

Murdoch, John E. (1982a) 'Ockham and the Logic of Infinity and Continuity', in Norman Kretzmann (ed.), *Infinity and Continuity in Ancient and Mediæval Thought*, Cornell University Press.

Murdoch, John E. (1982b), 'Infinity and Continuity', in Norman Kretzmann, Anthony Kenny and Jan Pinborg (eds), *The Cambridge History of later Mediæval Philosophy*, Cambridge University Press.

Murphey, Murray (1967), 'Peirce', in Paul Edwards (ed.), *The Encyclopædia of Philosophy*, London, Macmillan.

Nagel, Ernst and James R. Newman (1959) *Gödel's Proof*, London, Routledge & Kegan Paul.

Nagel, Thomas (1986) *The View From Nowhere*, Oxford University Press.

Nehamas, Alexander (1985) *Nietzsche: Life as Literature*, Harvard University Press.

von Neumann, John (1967) 'On the Introduction of Transfinite Numbers', trans. Jean van Heijenoort, in Jean van Heijenoort (ed.), *From Frege to Gödel: A Source Book in Mathematical Logic, 1879–1931*, Harvard University Press.

Newton, Isaac (1736) *The Method of Fluxions and Infinite Series*, trans. John Colson, London.

Newton, Isaac (1744) *Two Treatises of the Quadrature of Curves, and Analysis by Equations of an Infinite Number of Terms*, explained by John Stewart, London.

Newton, Isaac (1779–85) *Opera Quae Exstant Omnia*, ed. Samuel Horsley, London.

Newton, Isaac (1946) *Mathematical Principles of Natural Philosophy and System of the World*, trans. A. Motte and F. Cajori, Berkeley.

Newton-Smith, W.H. (1980) *The Structure of Time*, London, Routledge & Kegan Paul.

Nicholas of Cusa (1954) *Of Learned Ignorance*, trans. G. Heron, London, Routledge & Kegan Paul.

Nietzsche, Friedrich (1961) *Thus Spoke Zarathustra*, trans. R.J. Hollingdale, Harmondsworth, Penguin.

Nietzsche, Friedrich (1967) *The Will to Power*, trans. Walter Kaufmann and R. J. Hollingdale, New York, Random House.

Nietzsche, Friedrich (1974) *The Gay Science*, trans. Walter Kaufmann, New York, Random House.

Owen, G. E. L. (1957–8) 'Zeno and the Mathematicians', in *Proceedings of the Aristotelian Society*, 58.

Owen, H. P. (1967) 'Infinity in Theology and Metaphysics', in Paul Edwards (ed.), *The Encyclopædia of Philosophy*, London, Macmillan.

Parsons, Charles (1967) 'Foundations of Mathematics', in Paul Edwards (ed.), *The Encyclopædia of Philosophy*, London, Macmillan.

Parsons, Charles (1983) *Mathematics in Philosophy*, Cornell University Press.

Pascal, Blaise (1966) *Pensées*, trans. A.J. Krailsheimer, Harmondsworth, Penguin.

Peano, Giuseppe (1967) 'The Principles of Arithmetic, Presented by a New Method', trans. Jean van Heijenoort, in Jean van Heijenoort (ed.), *From Frege to Gödel: A Source Book in Mathematical Logic, 1879–1931*, Harvard University Press.

Pedersen, Kirsti Møller (1980) 'Techniques of the Calculus, 1630–1660', in I. Grattan-Guiness (ed.), *From the Calculus to Set Theory, 1630–1910*, London, Duckworth.

Peirce, C.S. (1891–2) 'The Law of Mind', in *The Monist*, 2.

Peirce, C.S. (1976) *The New Elements of Mathematics*, ed. C. Eisele, The Hague, Mouton.

The Infinite

Percival, Philip (1986) *Infinity, Knowability and Understanding*, unpublished Cambridge University Ph.D. thesis.

Plato (1961) *The Collected Dialogues*, ed. Edith Hamilton and Huntington Cairns, Princeton University Press.

Plotinus (1956) *The Enneads*, trans. Stephen MacKenna, London, Faber & Faber.

Priestley, W. M. (1979) *Calculus: An Historical Approach*, New York, Springer-Verlag.

Putnam, Hilary (1983) 'Models and Reality', in Paul Benacerraf and Hilary Putnam (eds), *Philosophy of Mathematics*, Cambridge University Press.

Quine, W. V. (1960) *Word and Object*, Cambridge, MIT.

Quine, W. V. (1964) 'Ontological Reduction and the World of Numbers', in *Journal of Philosophy*, 61.

Quine, W. V. (1969a) *Set Theory and its Logic*, Harvard University Press.

Quine, W. V. (1969b) *Ontological Relativity and Other Essays*, Columbia University Press.

Quine, W. V. (1974) *Methods of Logic*, London, Routledge & Kegan Paul.

Rescher, Nicholas (1955) 'Leibniz' Conception of Quantity, Number and Infinity', in *Philosophical Review*, 64.

Resnik, Michael (1966) 'On Skolem's Paradox', in *Journal of Philosophy*, 63.

Resnik, Michael (1969) 'More on Skolem's Paradox', in *Noûs*, 3.

Resnik, Michael (1981) 'Mathematics as a Science of Patterns: Ontology and Reference', in *Noûs*, 15.

Robinson, Abraham (1967) 'The Metaphysics of the Calculus', in Imre Lakatos (ed.), *Problems in the Philosopy of Mathematics*, Amsterdam, North Holland.

Robinson, G. (1964) 'Infinity', in *Proceedings of the Aristotelian Society*, Supp. Vol. 38.

Ross, David (1964) *Aristotle*, London, Methuen.

Rosser, J. Barkley (1936) 'Extensions of Some Theorems of Gödel and Church', in *Journal of Symbolic Logic*, 1.

Royce, Josiah (1900) *The World and the Individual*, London, Macmillan.

Rucker, Rudy (1982) *Infinity and the Mind: The Science and Philosophy of the Infinite*, Sussex, Harvester.

Russell, Bertrand (1900) *A Critical Exposition of the Philosophy of Leibniz*, Cambridge University Press.

Russell, Bertrand (1919) *Introduction to Mathematical Philosophy*, London, Allen & Unwin.

Russell, Bertrand (1926) *Our Knowledge of the External World as a Field for Scientific Method in Philosophy*, London, Allen & Unwin.

Russell, Bertrand (1935–6) 'The Limits of Empiricism', in *Proceedings of the Aristotelian Society*, 36.

Russell, Bertrand (1937) *The Principles of Mathematics*, London, Allen & Unwin.

Russell, Bertrand (1946) *History of Western Philosophy, and its Connection with Political and Social Circumstances from the Earliest Times to the Present Day*, London, Allen & Unwin.

Russell, Bertrand (1958) 'Mathematical Infinity', in *Mind*, 67.

Russell, Bertrand (1967) 'Letter to Frege', in Jean van Heijenoort (ed.), *From Frege to Gödel: A Source Book in Mathematical Logic, 1879–1931*, Harvard University Press.

Russell, Bertrand (1978) *Wisdom of the West, a Historical Survey of Western Philosophy in its Social and Political Setting*, London, Macdonald & Jane's.

Ryle, Gilbert (1954) *Dilemmas*, Cambridge University Press.

Salmon, W. C. (ed.) (1970) *Zeno's Paradoxes*, Indianapolis, Bobbs-Merrill.

Bibliography

Sanford, D. H. (1975) 'Infinity and Vagueness', in *Philosophical Review*, 84.

Sartre, Jean-Paul (1957) *Being and Nothingness: An Essay on Phenomenological Ontology*, trans. H. E. Barnes, London, Methuen.

Sayre, Kenneth M. (1983) *Plato's Late Ontology: A Riddle Resolved*, Princeton University Press.

Schopenhauer, Arthur (1969) *The World as Will and Representation*, trans. E. F. J. Payne, New York, Dover.

Schopenhauer, Arthur (1970) *Essays and Aphorisms*, trans. R. J. Hollingdale, Harmondsworth, Penguin.

Seligman, Paul (1962) *The 'Apeiron' of Anaximander: A Study in the Origin and Function of Metaphysical Ideas*, London, Athlone.

Shanker, S. G. (1987) *Wittgenstein and the Turning-Point in the Philosophy of Mathematics*, London, Croom Helm.

Simplicius (1882–1909) 'Commentary on Aristotle's *Physics*', in Hermann Diels (ed.), *Commentaria in Aristotelem Græca*, Berlin, Vols. X–XI.

Singer, D. W. (1950) *Giordano Bruno: His Life and Thought*, New York.

Skolem, Thoralf (1967a) 'Logico-Combinatorial Investigations in the Satisfiability or Provability of Mathematical Propositions: A Simplified Version of a Theorem by L. Löwenheim and Generalizations of the Theorem', trans. Stefan Bauer-Mengelberg, in Jean van Heijenoort (ed.), *From Frege to Gödel: A Source Book in Mathematical Logic, 1879–1931*, Harvard University Press.

Skolem, Thoralf (1967b) 'Some Remarks on Axiomatized Set Theory', trans. Stefan Bauer-Mengelberg, in Jean van Heijenoort (ed.), *From Frege to Gödel*, as above.

Smullyan, Raymond M. (1967) 'Continuum Problem', in Paul Edwards (ed.), *The Encyclopædia of Philosophy*, London, Macmillan.

Sorabji, Richard (1983) *Time, Creation and the Continuum: Theories in Antiquity and the Early Middle Ages*, London, Duckworth.

Spinoza, Benedictus de (1910) *Ethics*, trans. Andrew Boyle, London, Dent.

Strawson, P. F. (1966) *The Bounds of Sense: An Essay on Kant's 'Critique of Pure Reason'*, London, Methuen.

Tarski, Alfred (1959) 'What is Elementary Geometry?' in L. Henkin, P. Suppes, and A. Tarski (eds), *The Axiomatic Method, with Special Reference to Geometry and Physics*, Amsterdam, North Holland.

Taylor, A. E. (1903) *Elements of Metaphysics*, London, Methuen.

Taylor, Charles (1975) *Hegel*, Cambridge University Press.

Thomas, William (1968) 'Platonism and the Skolem Paradox', in *Analysis*, 28.

Thomas, William (1971) 'On Behalf of the Skolemite', in *Analysis*, 31.

Thomson, Garrett (1987) *Needs*, London, Routledge & Kegan Paul.

Thomson, James (1954) 'Tasks and Super-Tasks', in *Analysis*, 15.

Thomson, James (1967) 'Infinity in Mathematics and Logic', in Paul Edwards (ed.), *The Encyclopædia of Philosophy*, London, Macmillan.

Tiles, Mary (1989) *The Philosophy of Set Theory: An Historical Introduction to Cantor's Paradise*, Oxford, Blackwell.

Vlastos, Gregory (1967) 'Zeno', in Paul Edwards (ed.), *The Encyclopædia of Philosophy*, London, Macmillan.

Waismann, Friedrich (1982) *Lectures on the Philosophy of Mathematics*, ed. Wolfgang Grassl, Amsterdam, Rodopi.

Wang, Hao (1974) *From Mathematics to Philosophy*, New York, Humanities Press.

Warnock, G. J. (1987) 'Kant', in Magee, Bryan (ed.), *The Great Philosophers*, London BBC Books.

Webb, Judson C. (1980) *Mechanism, Mentalism and Metamathematics: An Essay on Finitism*, Dordrecht, Reidel.

Weyl, Hermann (1949) *Philosophy of Mathematics and Natural Science*, trans. O. Helmer, Princeton University Press.

Weyl, Hermann (1967) 'Comments on Hilbert's Second Lecture on the Foundation of Mathematics', trans. Stefan Bauer-Mengelberg and Dagfinn Føllesdal, in Jean van Heijenoort (ed.), *From Frege to Gödel: A Source Book in Mathematical Logic, 1879–1931*, Harvard University Press.

Whitehead, Alfred North and Bertrand Russell (1927) *Principia Mathematica*, Cambridge University Press.

Williams, Bernard (1973) *Problems of the Self*, Cambridge University Press.

Williams, Bernard (1981) *Moral Luck*, Cambridge University Press.

Wittgenstein, Ludwig (1961) *Tractatus Logico-Philosophicus*, trans. D. F. Pears and B. F. McGuinness, London, Routledge & Kegan Paul.

Wittgenstein, Ludwig (1969) *The Blue and Brown Books: Preliminary Studies for the 'Philosophical Investigations'*, Oxford, Blackwell.

Wittgenstein, Ludwig (1974a) *Philosophical Grammar*, ed. Rush Rhees and trans. Anthony Kenny, Oxford, Blackwell.

Wittgenstein, Ludwig (1974b) *Philosophical Investigations*, trans. G. E. M. Anscombe, Oxford, Blackwell.

Wittgenstein, Ludwig (1975) *Philosophical Remarks*, ed. Rush Rhees and trans. Raymond Hargreaves and Roger White, Oxford, Blackwell.

Wittgenstein, Ludwig (1976) *Lectures on the Foundations of Mathematics*, ed. Cora Diamond, Sussex, Harvester.

Wittgenstein, Ludwig (1978) *Remarks on the Foundations of Mathematics*, ed. G. H. von Wright, R. Rhees, and G. E. M. Anscombe and trans. G. E. M. Anscombe, Oxford, Blackwell.

Wittgenstein, Ludwig (1979) *Notebooks: 1914–1918*, ed. G. H. von Wright and G. E. M. Anscombe and trans. G. E. M. Anscombe, Oxford, Blackwell.

Wright, Crispin (1979) *Wittgenstein on the Foundations of Mathematics*, London, Duckworth.

Wright, Crispin (1982) 'Strict Finitism', in *Synthèse*, 51.

Wright, Crispin (1985) 'Skolem and the Skeptic', in *Proceedings of the Aristotelian Society*, Supp. Vol. 59.

Wrigley, Michael (1987) *Wittgenstein's Early Philosophy of Mathematics*, unpublished University of California Ph.D. thesis.

Zermelo, Ernst (1967) 'Investigations in the Foundations of Set Theory I', trans. Stefan Bauer-Mengelberg, in Jean van Heijenoort (ed.), *From Frege to Gödel: A Source Book in Mathematical Logic, 1879–1931*, Harvard University Press.

Index

absolute continuity 157

absoluteness 1, 55–6, 78, 101–2, 104, 127–8, 188, 219, 230

actual infinite, the: as an amalgam of the mathematically infinite and the metaphysically infinite 44, 77, 103, 113; defined 39–40; in relation to the potential infinite 39–40, 47, 51–2, 77, 93, 117, 132, 142, 149, 171, 204, 208

'all at once' 29, 39–40, 43–4, 47, 81, 88, 110

analogy of the field of vision, the 189–94 *passim*, 226

analysis 58, 64, 66, 68, 114, 173; non-standard *see* non-standard analysis; *see also* analytic geometry

analytic geometry 58, 62, 66, 68, 114

Anaxagoras 34, 36

Anaximander 15, 17–27 *passim*, 34–8, 99, 106, 202

antinomies, Kant's *see* Kant's antinomies

apeiron, to 15, 17–28 *passim*, 35, 99, 187, 202, 237; the imposition of the *peras* on it 19–20, 27, 105, 246 n.16

appearance/reality distinction, the 18, 23–28 *passim*, 35, 45, 77–9, 85, 92–104, *passim*; in relation to the mathematical/metaphysical distinction 24, 45–6, 78–9, 87–8, 97–9, 222

Aquinas, St T. 48–9, 78, 184

Archimedes 30, 32–3, 51, 55, 57, 61–2, 66

Archytas 28–9, 38, 56, 80, 134

Aristotle 5, 15, 29, 34–44, 70, 78, 82, 84, 88, 92, 96–7, 99, 104, 110, 112, 138, 143, 198, 207, 211, 213, 216, 222, 237

n.2, 237 n.5, 240 n.23; and the actual/potential distinction 39–44, 47, 52, 73, 143, 204, 207–8, 216; on the infinitude of time 38–40, 44, 47, 88, 91, 207–8, 216, 221; his influence on his successors 34, 39–40, 44–56, 76–81 *passim*, 87, 90–3, 99–100, 110, 113, 117, 132–3, 208; on his predecessors 17–18, 34–9, 43–4, 75, 80, 97, 99, 237 n.5; on Zeno's paradoxes 3, 5, 25, 37–8, 42, 70–3 *passim*, 104, 158, 216

arithmetic 29, 114, 133, 153, 174–85 *passim*

'as if' 89, 95, 108, 135, 186, 223–4, 228, 232

atomism 25–7, 35–7, 79, 82, 92, 104; logical *see* logical atomism

Augustine, St 46–51 *passim*, 78

Aureol, P. 52

Avicenna 48

axiomatization 29–30, 75, 77, 116, 129–30, 134, 154–5, 160, 169, 172–83, 243 n.44; *see also* the Euclidean paradigm

axioms of infinity (axioms concerning the infinite) 116, 129, 155, 157, 166, 179, 182

Beethoven, L. 145

Benardete, J. 5, 9, 70, 143

Beresford, G. 239 n.16

Bergson, H. 103–4, 107, 110, 117, 158, 203

Berkeley, G. 57, 65–6, 80–2

Boethius, A. 47

Bolzano, B. 66, 112–17, 198, 242 n.4

Bradley, F. 101–3, 187

261

Index

Empedocles 108
empiricists, the 75, 80–5, 88, 90, 143, 205–6, 222
eternal recurrence 108–9
eternity 8, 34, 39, 44, 75, 77; as timelessness 23, 28, 34, 47, 77, 194, 226, 232; *see also* whether time is infinite
Euclid 29, 75, 114, 173, 199, 243 n.48
Euclidean paradigm, the 29–30, 114, 129–30, 172, 182, 243 n.48; *see also* axiomatization
Eudoxus, 30, 32, 51, 55, 57, 61, 66
Euler, L. 66
excluded middle, the law of the *see* the law of the excluded middle
exhaustion, the method of *see* the method of exhaustion
existentialists, the 100, 105–8, 194, 203–4, 229

façon de parler 51, 64, 68, 79, 134, 136, 170–1, 186, 206
faith 56, 94, 101, 105–6, 117, 232
femaleness 19, 202
Fermat, P. 60, 63
Ficker, L. 189
finitism 134–6, 179; *see also* Hilbert's programme
Foucher, S. 78
Frænkel, A. 129
frameworks, infinite *see* infinite frameworks
freedom 84–6, 92–7 *passim*, 105–7, 195
Frege, G. 114–16, 121, 169, 242 n.21

Galileo 54–5
Galileo's paradox 54
Gassendi, P. 76–7, 81
geometry 29, 38, 43, 58, 64, 66–9 *passim*, 75, 82, 114, 132, 172–3, 243 n.48; analytic *see* analytic geometry
Ghazālī, A. 48–9
God 47–56 *passim*, 76–9, 84, 86, 92, 94, 105, 112, 121, 128, 189, 240 n.12, 248 n.3; identified with the infinite 46, 55–6, 77–8, 98, 101
Gödel, K. 104, 129–30, 136, 140, 154–9 *passim*, 172–81 *passim*
Gödel's theorem 129–30, 136, 154, 172–85, 197, 202, 246 n.8
Goldbach, C. 133
Goldbach's conjecture 133, 174

'grammar' of 'infinity', the 52, 138–40, 195, 198, 206, 214–17 *passim*, 221; *see also* the Wittgensteinian critique
Grandi, G. 72
Gregory of Rimini 51–4, 110–11, 113, 118

von Haller, A. 96
Halley, E. 238 n.7
Hamilton, W. 100, 203
Haussdorf, F. 157
Hawking, S. 220
Heidegger, M. 106–7, 137, 203, 229, 240 n.9, 247 n.5
Hegel, G. 66, 96–101, 110, 112, 143, 147, 158, 188, 198, 203, 219, 222, 240 n. 17; his distinction between true infinity and spurious infinity 99–100, 117, 143, 188, 203; his influence on his successors 100–2, 112, 117, 187–8; on Kant 96–100 *passim*
Henry of Harclay 53
Herbart, J. 103
hierarchy of Sets, the 149–51, 157–8, 161–6, 170–1, 179–80, 182, 186, 200, 225; *see also* the iterative conception of sets; whether Sets form a determinate totality
Hilbert, D. 1, 9, 131–40 *passim*, 154, 178–9, 223, 246 n.8
Hilbert's programme 135–6, 178, 246 n.8; *see also* finitism
Hobbes, T. 80–1, 88, 98
hope 232
de l'Hôpital, G. 65–6
von Hügel, F. 101, 105–6
human finitude 11–13, 76, 96, 145, 190, 203, 218, 223–33 *passim*, 248 n.3; in early Greek thought 18–19; its fundamental nature xi, 86, 100, 106–7, 195, 218–29 *passim*; in Kant 84–6, 94–6, 106, 190, 204, 219, 233, 240 n.9; in medieval and Renaissance thought 45, 47, 53–6; in post-Kantian thought 98–107 *passim*, 194–5, 203, 219; as revealed in various lmitations ix–x, 45, 47, 53–6, 75, 80, 98, 101, 105–7, 142–3, 165–6, 170, 181, 186, 211–16 *passim*, 218–20, 226; in Wittgenstein 194; *see also* death
Hume, D. 75, 80–2, 115, 205
Husserl, E. 103

Index

mathematics: classical *see* classical mathematics; early Greek *see* early Greek mathematics; its foundations 9, 12, 114–17, 129–37; its methods and nature 75, 82, 114, 116, 132–42 *passim*, 151–5 *passim*, 161–2, 167–9, 172–4, 178–85, 211, 214, 216; as the science of the infinite 29, 73–4, 147, 199; transfinite *see* transfinite mathematics; whether it is a forum for the study of the infinite 29, 73–4, 110–22 *passim*, 127–9, 131–9 *passim*, 147, 171, 179, 198–9, 201, 242 n.1; *see also* set theory

matter 18, 35–9 *passim*, 46, 77–8; whether it is infinitely divisible 5, 26–7, 37–42 *passim*, 79, 82, 89–93 *passim*, 134, 207–8, *see also* atomism, whether space is infinitely divisible

meaning 132–43 *passim*, 160–70, 172, 182–5, 187–97 *passim*, 214; its infinitude 183–5, 192, 196, 217; and use 183–4; as value (the meaning of life) 107, 189, 194–5, 226–33 *passim*; *see also* understanding

measuring infinite sets 122–3, 127, 139, 152–6; *see also* comparisons of size between infinite sets

medievals, the 45–56, 68, 75, 77, 90, 92, 110, 134, 139, 208

Melissus 23–5, 43, 48

metaphysical conceptions of the infinite *see* mathematical and metaphysical conceptions of the infinite

metaphysically infinite as the (limited) whole, the 1–2, 24, 43, 46, 55, 77–8, 86, 88, 97, 158, 188, 193–200 *passim*, 220–4 *passim*, 231

method of exhaustion, the 30, 55, 57, 61, 66

Murdoch, I. 219

mysticism 46, 56, 77, 101, 140, 143, 189, 193, 203, 247 n.24

Nagel, E. 177

Nagel, T. 227

natural numbers: as cardinals 152; how they compare in size, as a set, with the set of real numbers 120–2, 128, 132, 136, 140, 154, 156, *see also* Cantor's unanswered question, the continuum hypothesis; defined 5, 151; their endlessness 10–11, 29,

37–42 *passim*, 81, 125, 133, 137, 156, 199, 206–16 *passim*; as ordinals 125–7, 149, 151; whether they form a determinate totality 10–11, 46–7, 102, 112, 117, 126–8, 132–6 *passim*, 141, 201–2, 209, 213

Neoplatonism 45–9, 77–8, 101, 203

von Neumann, J. 151

Newman, J. 177

Newton, I. 30, 63–5, 79, 84, 239 n.9; *see also* Newtonians

Newtonians 64, 79, 81, 92

Nicholas of Cusa 45, 55–6, 77–8, 97–9, 102, 203, 240 n.12

Nietzsche, F. 96, 108–9, 232–3

non-standard analysis 69

nothingness 227–8

number-predicates 175–6

numbers: cardinal *see* cardinal numbers; infinite *see* infinite numbers; natural *see* natural numbers; ordinal *see* ordinal numbers; rational *see* rational numbers; real *see* real numbers

ordinal numbers 122–7, 149–53, 160–1, 164, 210–11; their endlessness 125–7, 149, 152, 156–7, 161, 164, 206–12 *passim*, 244–5 n.7; *see also* Ω

Oresme, N. 60

otherness ix, 11, 97, 100–2, 202, 218–19, 226, 230–2

paradox of Achilles and the tortoise, the 3–4, 25, 42

paradox of the arrow, the 25, 42–3, 70–1, 158

paradox, the Burali-Forti *see* the Burali-Forti paradox

paradox of the divided stick, the 5, 26, 138, 214

paradox of the even numbers, the 7, 54, 79, 111, 113–14, 118

paradox, Galileo's *see* Galileo's paradox

paradox of the gods, the 4–5, 143

paradox of the hotel, the 9

paradox of the lamp, the 70–1, 239 n.16

paradox of the pairs, the 7–8, 118, 156

paradox of the runner, the 25, 42, 71, 103–4

paradox, Russell's *see* Russell's paradox

paradox of the Set of all Sets, the 149–51, 170, 182, 197

Index